EVOLUTIONARY PHYSIOLOGY AND BIOCHEMISTRY - ADVANCES AND PERSPECTIVES

Evolutionary Physiology and Biochemistry - Advances and Perspectives
http://dx.doi.org/10.5772/64635

Contributors
Vladimir F. Levchenko, Olga S. Alekseeva, Natalia F. Avrova, Margarita G. Belekhova, Galina B. Belostotskaya, Ivan T. Demchenko, Kira V. Derkach, Ludmila E. Dmitrieva, Nadezhda M. Dubrovskaya, Michael M. Galagudza, Elizaveta I. Galperina, Tatyana A. Golovanova, Diana M. Guillemard (Tsaparina), David M. Holloway, Ekaterina E. Kats, Alexander B. Kazansky, Natalia B. Kenigfest, Alexander N. Knyazev, Alexander I. Krivchenko, Olga V. Kruchinina, Sergey V. Kuznetsov, Natalia N. Kuznetsova, Natalia N. Nalivaeva, Irina V. Nerubatskaya, Ekaterina A. Panasevich, Natalya P. Prutskova, Marat A. Sabirov, Gennady A. Savostyanov, Elena V. Seliverstova, Alexandr N. Shepovalnikov, Alexander O. Shpakov, Vladimir A. Sizonov, Alexander V. Spirov, Maksim A. Terpilowski, Denis B. Tikhonov, Olga P. Timofeeva, Michail N. Tsitseroshin, Natalia L Tumanova, Dmitrii S. Vasilev, Natalia D. Vdovichenko, Yulia A. Vlasova, Larisa G. Zaytseva, Boris S. Zhorov and Igor A. Zhuravin

Published by InTech
Janeza Trdine 9, 51000 Rijeka, Croatia

Publishing Process Manager Iva Simcic
Technical Editor SPi Global
Cover InTech Design team

First published February, 2018
Printed in Croatia
Legal deposit, Croatia: National and University Library in Zagreb

Additional hard copies can be obtained from orders@intechopen.com

Evolutionary Physiology and Biochemistry - Advances and Perspectives
p. cm.
Print ISBN 978-953-51-3857-0
Online ISBN 978-953-51-3858-7

For EU product safety concerns:
IN TECH d.o.o., Prolaz Marije Krucifikse Kozulić 3, 51000 Rijeka, Croatia,
info@intechopen.com or visit our website at intechopen.com.

Contents

Leon A. Orbeli

Chapter 1

Introduction: The 60th Anniversary of the Institute of Evolutionary Physiology and Biochemistry, Russian Academy of Sciences, Saint Petersburg

Alexander N. Knyazev

Additional information is available at the end of the chapter

http://dx.doi.org/10.5772/intechopen.73669

1. Introduction

The history of evolutionary physiology as an independent branch of general physiology goes back more than 100 years. In the USSR, until the first third of the twentieth century, the problems of evolutionary physiology were addressed mainly by individual scientists and small groups scattered in different research and academic institutions. In October 1950, one of such groups had emerged in Leningrad under the leadership of Academic Leon Abgarovich Orbeli; in September 1954, it was transformed into a small laboratory. Over many years, L.A. Orbeli sought to create in the country a large multidisciplinary research center that would be united by a common idea and focused entirely on evolutionary problems in all their diversity. The first attempt to establish such a center in Koltushi Village near Leningrad (Institute of Evolutionary Physiology and Pathology of Higher Nervous Activity) ended with failure—the Institute was liquidated. After all, thanks to Orbeli's great efforts, the Sechenov Institute of Evolutionary Physiology of the USSR Academy of Science (presently Sechenov Institute of Evolutionary Physiology and Biochemistry of the Russian Academy of Sciences, IEPB) had been founded in Leningrad – see **Figure 1**.

Soon after that, in March 1956, L.A. Orbeli articulated the main tasks and methods of evolutionary physiology at the First Session on Evolutionary Physiology, and this was actually a manifesto of the new branch of domestic physiology (historical article of L.A. Orbeli, see below). Four months later, in April 1956, the Presidium of the Russian Academy of Sciences approved the structure of the new Institute and its research plan for the coming years.

A first large-scale attempt to generalize the achievements and prospects of evolutionary physiology in collected reviews of its major divisions was made more than 30 years ago by

Figure 1. Sechenov Institute of Evolutionary Physiology and Biochemistry of the Russian Academy of Science at present.

Academic E.M. Kreps, who then headed the IEPB. He was an inspirer and editor-in-chief of the multi-volume series of books issued under the common title "Handbook of Physiology." Two volumes of the series were devoted to domestic and foreign achievements as well as major trends and prospects in evolutionary physiology (Evolutionary Physiology, Pt. 1, 2, *Handbook of Physiology,* E.M. Kreps, ed., 1979, 1983).

Thirty years after the publication of this Handbook, the present compendium offers the reviews of some aspects of evolutionary physiology that have been tackled for the last six decades and are still under study at the IEPB. The reviews are preceded by a verbatim report of that momentous Orbeli's speech, which became a Bible of domestic evolutionary physiology and retains its importance undiminished thus far. The content and rhetoric of this document can be regarded, on the one hand, as a monument to that epoch and, on the other hand, as a brilliant foresight of the developmental pathways and principles of evolutionary physiology for many years to come.

Unfortunately, the limited volume of the present publication does not allow us to cover the variety of evolutionary studies being successfully conducted at the IEPB. Such important areas such as evolutionary physiology of sensory and visceral systems, evolutionary immunology, endocrinology, neuroendocrinology, psychoneuroendocrinology, somnology, and some others have to be left beyond the scope of this book. The multiple evolutionary studies performed on invertebrate animals are also omitted. Without question, further publication of relevant reviews would be extremely useful for the popularization and promotion of evolutionary studies and evolutionary physiology in general.

The research status quo at the IEPB is presented on the site http://iephb.ru/ as well as in the Journal of Evolutionary Biochemistry and Physiology [ISSN: 0022-0930 (Print) 1608-3202 (Online)], http://link.springer.com/journal/volumesAndIssues/10893. The Journal was founded in 1965 by the then Director of the IEPB Academic. E.M. Kreps.

http://dx.doi.org/10.5772/intechopen.73669

Author details

Alexander N. Knyazev

Address all correspondence to: ank50@mail.ru

I.M. Sechenov Institute of Evolutionary Physiology and Biochemistry RAS, St. Petersburg, Russia

Chapter 2

Brain Gangliosides and Their Function as Natural Adaptogenes

Natalia F. Avrova and Yulia A. Vlasova

Additional information is available at the end of the chapter

http://dx.doi.org/10.5772/intechopen.73648

Abstract

In brain gangliosides and phospholipids of stenothermal cold-water teleost fishes, higher content of polyenoic and monoenoic fatty acids was revealed than in brain gangliosides and phospholipids of warm-water stenothermal teleosts. The changes in fatty acid composition of lipids during adaptation of fishes to living in cold water (or at great water depth) are directed to the maintenance of liquid-crystalline state of cell membranes and their optimal fluidity, physical state, and microheterogeneity. The results of cluster analysis of the data on composition of carbohydrate component of brain gangliosides of various ectothermic vertebrates were used to create the dendrogram. This dendrogram was found to correspond appreciably to the tree of classical taxonomy of vertebrates. The changes in molecular organization of brain gangliosides in the course of evolution of vertebrates are suggested to contribute to differentiation of brain and complication of its functions in phylogenesis. The main brain gangliosides (GM1, GD1a, GD1b, GT1b) may be considered to be typical adaptogens. They protect neurons against the action of excitatory amino acids, hydrogen peroxide, amyloid β-peptide, and other toxins. Protective effect of gangliosides against these toxins depends on activation of Trk receptor tyrosine kinase and downstream protein kinases.

Keywords: gangliosides, adaptogens, neuroprotection, signal transduction pathways

1. Introduction

From the beginning of 1960s of the previous century till the present time, the main investigations in the laboratory of comparative neurochemistry of our Institute (from 2014 becoming a part of laboratory of molecular endocrinology and neurochemistry) were devoted to studies of brain lipids. This choice was made by Professor E.M. Kreps, who founded the laboratory and was its head for many years. Professor Kreps was a prominent and outstanding scientist. He was a member of Russian Academy of Science (RAS) and for many years, a Head of Division of Physiology of Russian Academy of Sciences. Professor E.M. Kreps had a gift of scientific foresight, as in 1960s, studies of cell membrane lipids did not attract much attention from investigators. But later it became evident that such studies were necessary for understanding of biochemical mechanisms of action of hormones, second messengers, mediators, and other physiologically active compounds, of processes of animal adaptation to changes of environment and functional activities, and of mechanisms of various disease pathogenesis. The clinical trials of various lipids may result in appearance of new drugs. In the laboratory of comparative neurochemistry, the comparative studies of various lipids, including phospholipids, cholesterol and its esters, cerebrosides, sulphatides, and gangliosides, were performed [1–3]. The aim of the present short review is to describe as an example of comparative brain lipid investigations performed under the guidance of Professor E.M. Kreps the studies of vertebrate brain gangliosides and to characterize the recent data on the mechanism of exogenous ganglioside protective action on neurons and cells of neuronal cell lines obtained by his collaborators.

2. Changes in brain ganglioside fatty acid composition as a result of natural adaptations of fishes to the conditions of environment

Gangliosides are the most complex glycolipids in animals, containing sialic acids. The four main brain gangliosides in mammals (GM1, GD1a, GD1b и GT1b) consist of fatty acid, long-chain base residues and carbohydrate chain of four monosaccharides which form bonds with 1–3 sialic acid residues. Gangliosides appeared at relatively late stages of animal evolution, as components of cell membranes of various organs of Deuterostomia (types of Echinodermata and Chordata). It was shown that in the course of evolution of vertebrates, ganglioside content of brain increases along with the increase in the degree of brain differentiation leading to its more complex organization [3, 4]. Comparative study of cell membrane lipids of various organs appears to be a fruitful approach to elucidation of their role in adaptation of organisms to changing conditions of environment and functional activity [3, 5].

The differences in fatty acid composition of brain gangliosides were found to be highly significant when stenothermal teleosts living at low and relatively high water temperature were compared. These differences are more pronounced than the differences in fatty acid composition of phospholipids (**Figure 1**). Probably, it may be due to the fact that the main places of ganglioside localization in vertebrates are plasma membranes of neurons, including their synaptic membranes, the maintenance of their functional activity at optimal level being especially important to organisms [5]. Thus, saturated fatty acids content in brain ganglioside was much lower in cold-water stenothermal teleosts than in warm-water ones (83.4 ± 1.5 and 50.9 ± 4.9% from total fatty acid content, respectively). The content of monoenoic and polyenoic fatty acids in brain gangliosides was, on the contrary, much higher (**Figure 1**) in cold-water fishes (35.3 ± 3.6% and

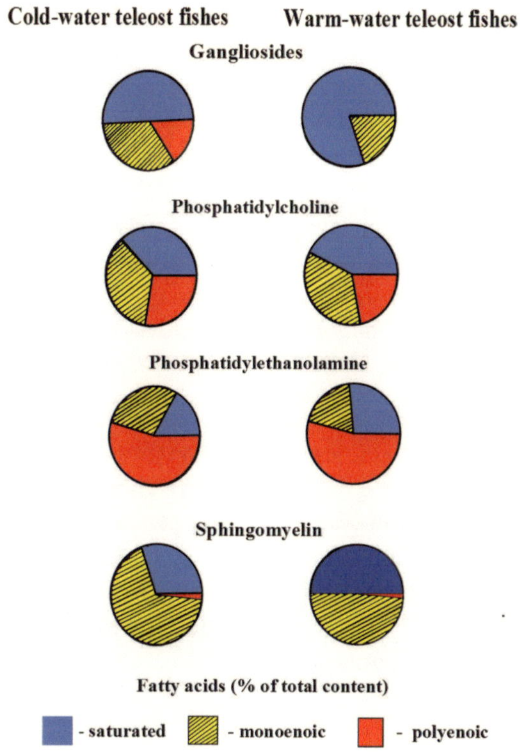

Figure 1. Fatty acid composition of lipids from brain of warm-water and cold-water stenothermal species of teleost fishes.

13.8 ± 4.0%, respectively), as compared with warm-water fishes (16.1 ± 1.6% and less than 1% from their total content, respectively). The differences between the two groups of fishes were highly significant ($p < 0.01$ in all cases).

Fatty acid composition of gangliosides and phospholipids from brain of seven stenothermal cold-water teleost species living at water temperature of 0–10°C (*Bathylagus antarcticus, Lampanyctus australis, Antimora rostrata, Coelorinchus* sp., *Comephorus baicalensis, Comephorus dybowskii, Cottocomephorus inermis*) and of seven stenothermal warm-water teleost species living at temperature of 23–25°C (*Cheilopogon exsilience, Lepophidium profundorum, Calamus* sp., *Coryphaena hippurus, Lethrinus chrysostomus, Rhomboplites aurorubens, Sphyraena picudilla*) is shown in **Figure 1**. The highest content of unsaturated fatty acids in brain gangliosides was characteristic of the fishes living in water at great depth. At adaptations of fishes to environmental temperature, the most pronounced changes was revealed in the contents of fatty acids 18:0, 22:1, 24:1, and 22:6 ω3 in brain phospholipids and gangliosides, and they may be called "the adaptation tools" [3].

At the natural adaptations of cartilaginous and ganoid fishes to the temperature of the habitat, the changes in the content of saturated and monoenoic fatty acids in brain gangliosides were revealed, while polyenoic fatty acids were practically absent in these brain lipids. Studying 37 species of cartilaginous, ganoid, and teleost fishes it was found [6] that the portion of saturated fatty acids in brain gangliosides increased essentially and significantly at augmentation of the environmental temperature ($r^2 = 0.21$, $r = 0.46$, $p < 0.005$). In contrast, the contents of monoenoic and long-chain fatty acids in fish brain gangliosides correlated negatively to the environmental temperature ($r^2 = 0.26$, $r = -0.51$, $p < 0.002$ and $r^2 = 0.34$, $r = -0.58$, $p < 0.001$, accordingly), but the correlation between the content of polyenoic fatty acids in brain gangliosides of teleost, ganoid, and cartilaginous fishes and the temperature of their habitat was not revealed [6]. It may be explained by the fact that polyenoic fatty acids are characteristic mainly for gangliosides from teleost brain but not for gangliosides from ganoid and cartilaginous fish brain [3, 7].

The introduction of one double bond in the fatty acid molecule decreases the temperature of its melting point by several tens of degrees, and the introduction of several double bonds in the molecule has more pronounced effect. The differences in brain ganglioside fatty acid composition of warm-water and cold-water species appear to be the results of idioadaptations of fishes to the temperature of their habitat. They are directed to the maintenance of brain cell membrane fluidity, physical state, and microheterogeneity at the optimal level for function of enzymes, receptors, and other bioactive compounds and make possible the survival of animals in changing conditions of environment.

3. The differences in the structure and composition of carbohydrate component of ganglioside molecule between representatives of various vertebrate classes

We did not reveal the correlation between the composition and structure of brain ganglioside carbohydrate component and the water temperature at which the fishes are living. It was of interest to apply one of the cluster methods of analysis which may provide objective interpretation of a large amount of experimental data [6]. To estimate the similarity or the difference of various species, a number of features of their organization may be used which may be determined quantitatively. We have not met in the literature such studies devoted to brain lipids. In our case, the construction of dendrograms was carried out by the unweighted pair-group method of averages. "Unweighted" means that each of the parameters has equal significance or "weight".

The content of individual gangliosides (GP, GQ, GT1b, GD1b, GD1a, GD3, GM1) and the summarized content of their two minor fractions (GM2 + GM3) in brains of 24 species of exothermic vertebrates, investigated by us, were used as the eight parameters of the molecular organization, capable of being evaluated quantitatively. (To characterize ganglioside structure we used the nomenclature of Svennerholm, which is the most widely used nomenclature of gangliosides. Letters M, D, T, Q, and P indicate the number of sialic acid residues in the ganglioside molecule mono, di, tri, tetra, and pentasialogangliosides, respectively. Thus, GM1 is monosialoganglioside. The figure (1) corresponds to four saccharide residues, (2) corresponds to three saccharide residues, and (3) corresponds to two saccharide residues in the ganglioside molecule. These parameters were evaluated as percentage of the total content of brain gangliosides of the species and were expressed in arbitrary units. The content of the ganglioside was expressed as value 1 if its content in the brain of this species is equal to 1–10%, as value 2 at 11–20%, as value 3 at 21–40%, as value 4 if the content was more than 40%. If the content of ganglioside fractions in the brain of a species was less than 0.5% of total brain gangliosides, it was labeled by the 0 value. Then, the dendrograms were generated by PAUP program ver.4.0b8 for Mackintosh using the unweighted pair-group method of averages optimized according to the maximum parsimony principle [6].

The dendrogram, constructed based on the cluster analysis of the data on composition of carbohydrate component of brain gangliosides of the ectothermic vertebrates (**Figure 2**), corresponds appreciably to the tree of classical taxonomy of vertebrates. The species of cartilaginous and ganoid fishes form separate clusters and are sisterly branches for each other. Species of these fishes do not form sisterly groups with any representatives of teleosts, amphibians, or reptilians. All teleost fishes are in ranges of single separate branches on the cladogram. But in contrast to the classical systematics of the vertebrates, two investigated species of amphibians (of Anura group) do not form a separate branch but stand among species of teleosts forming sisterly groups with their species. Probably, these animals have the composition of brain ganglioside carbohydrate component characteristic of their common ancestors. It should also be noted that there is a great difference in the level of central nervous system organization between various species of teleost fishes. It is well known to the zoologist that some species of teleosts have a more differentiated and higher organized brain than representatives of the class of Amphibia. If the level of brain organization has an influence on the structure and

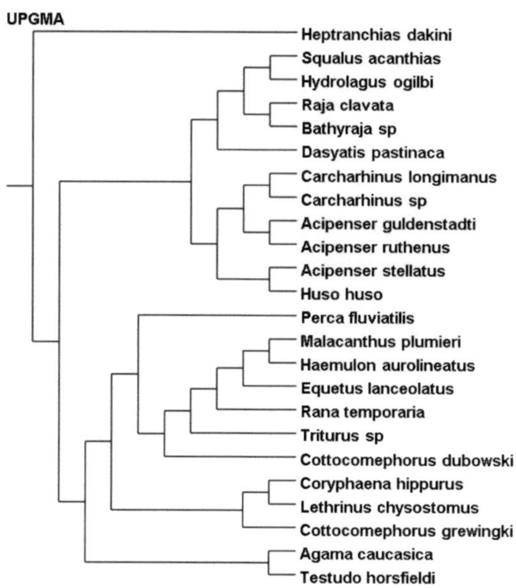

Figure 2. The dendrogram of relative similarity of different vertebrate species according to the parameters of the molecular organization of the carbohydrate component of the brain gangliosides.

composition of brain ganglioside carbohydrate component, it may explain why Anura do not form a separate branch on the dendrogram. In contrast to amphibians, the investigated reptilians form a separate branch on the dendrogram (**Figure 2**).

Among the higher vertebrates, only reptilians are submitted on the cladogram. But much lower content of polysialogangliosides and much higher portion of monosialogangliosides are characteristic of both reptilian and avian and mammalian brain in comparison with the lower vertebrate brain. As it is known, there were cardinal modifications in the organization of animals and their separate organs at entrance of vertebrates on dry land, which are a typical example of an aromorphosis [8]. Carbohydrate components of the glycolipids and glycoproteins play an essential role in cell differentiation and intercellular interaction, including cell recognition and adhesion, especially in the development of organs and tissues. Modifications of the brain gangliosides composition and structure, associated with appearance of the higher vertebrates, have provided, apparently, together with many other biochemical modifications, the molecular bases of the aromorphosis at transferring of vertebrates to the terrestrial lifestyle and amniotic development of embryos, that was accompanied by differentiation and complication of brain functions.

Other differences in composition of the brain gangliosides in various vertebrate classes are observed in the comparison of bony and cartilaginous fishes. They, in particular, reflect elongation of the carbohydrate chain of these lipids during evolution of vertebrates; it is typical also of ontogenetic development of mammals. These modifications of brain gangliosides composition in vertebrates may be also suggested to be associated with aromorphic modifications in the organization of animals.

In our research, the ganoid fishes have formed a single cluster; simultaneously, they are included into sisterly group with high-organized cartilaginous fishes. Views of zoologists on systematic position of the ganoid fishes are inconsistent. These fishes are surveyed as one of the superorders of the class of bony fishes. But some zoologists, on the contrary, emphasize a common origin of ganoids and cartilaginous fishes (e.g., see [9]. Our results (**Figure 2**) correspond with the data of the latter group of morphologists. Researches of hybridization of DNA of various fish species [10] and the study of composition and structure of various lipids (phospholipids, gangliosides, cerebrosides, and sulphatides) from fish brain [3] provide evidence that ganoid fishes have essential difference both from teleosts and from cartilaginous fishes that it is necessary to consider them as the taxon of the higher order than one of the superorders of bony fishes, probably, as a separate class or subclass.

4. Protective effect of main brain gangliosides (GM1, GD1a, GD1b, and GT1b) on damaged nerve cells or cells of neuronal cell lines

Gangliosides possess the functions of adaptogenes not only in ectothermic animals but in mammals too. It appears not to be due to the changes of brain ganglioside fatty acid composition, as the relative content of their saturated and unsaturated fatty acids was shown not to change as a result of cold stress [11]. But exogenous gangliosides administered to mammals with damaged brain increase the viability of brain neurons and improve the functional state

http://dx.doi.org/10.5772/intechopen.73648

of the animals, if the injury is caused by ischemia and reperfusion, various toxins, or trauma. Brain gangliosides possess the protective action on cultured neurons as well. The protective effect is characteristic for the main brain gangliosides (GM1, GD1a, GD1b, and GT1b). The most stable ganglioside GM1 is usually used in such experiments. It was found that the protective effect of GM1 against the toxic action on the nerve cells of serum-free medium devoid of growth factors or glutamate depended on the activation of Trk receptor tyrosine kinase [12–14]. We have for the first time shown that GM1 ganglioside increases the viability of nerve cells or cells of neuronal lines at application of other toxic compounds—of amyloid peptides [15–17] or hydrogen peroxide [14], the protective action of GM1 being also based on the activation of Trk receptor tyrosine kinase.

5. Gangliosides exert the protective effect on neurons and cells of neuronal cell lines not only at micromolar concentrations but at nanomolar concentrations as well

The protective effect of gangliosides on cultured nerve cells is, as a rule, studied using their micromolar concentrations (10–50 μM). But in cerebrospinal fluid (CSF) and in brain intercellular space of humans and animals, gangliosides are present in nanomolar concentrations. Thus, the total content of four main brain gangliosides (GM1, GD1a, GD1b, and GT1b) was found to constitute on average 92 nM in human CSF [18]. The physiological concentrations of gangliosides which act on brain nerve cells from outside in vivo appear to be nanomolar concentrations. We were the first to show that the protective effect of gangliosides against the toxic action of glutamate on cerebellar granule cells is well expressed both in micro and at nanomolar concentrations [19]. Thus, glutamate increased the number of dead granule cells from 12 ± 3% to 47 ± 4% of total cell number. But if granule cells were preincubated with 10 nM or 10 μM GM1 prior to application of glutamate, then the number of dead neurons decreased to 24 ± 4% and to 20 ± 5%, respectively ($p < 0.01$ in all cases). Gangliosides GD1a, GD1b, and GT1b at nanomolar concentrations also increased granule cells' viability. Later on [14], it was shown by us that the protective effect of GM1 at nanomolar concentrations also depended on activation of Trk receptor tyrosine kinase, as it was previously shown for gangliosides at micromolar concentration.

GM1 and GD1a gangliosides at nanomolar concentrations increase the viability of the cells of neuronal line PC12 exposed to hydrogen peroxide as well, their effect being also mediated by activation of Trk receptor tyrosine kinase. But in these cells, the protective effect of gangliosides at nanomolar concentrations was lower than their effect at micromolar concentrations [20].

6. The mechanism of neuroprotective effect of GM1 and other main brain gangliosides

Activation by GM1 ganglioside of protein kinase regulated by extracellular signals (ERK1/2) and of protein kinase B (Akt) takes place downstream of Trk receptor tyrosine kinase; it was shown using brain slices [21] and PC12 cells [20]. Activation of these protein kinases by GM1 is of importance for realization of the protective effect of GM1. Thus, it was shown [20] that the protective effect of GM1 against the toxic action of hydrogen peroxide is significantly diminished in the presence of the inhibitor of ERK1/2, or Akt, or protein kinase C. These data are in agreement with the data showing that ERK1/2 activation by GM1 increases the viability of retinal neurons after the axotomy of optic nerve [22]. But only in the presence of the inhibitors of all these protein kinases, the diminution of the protective effect of GM1 is pronounced and comparable with the effect of the inhibitor of Trk receptor tyrosine kinase which abolishes the protection.

Using immunoblotting it was shown that GM1 ganglioside both at nano and micromolar concentrations activated protein kinase B (Akt) in control PC12 cells [20]. The effect of GM1 on ERK1/2 activity was shown to be more pronounced in micro than in nanomolar concentration. Hydrogen peroxide itself activated ERK1/2. And preincubation of PC12 cells with GM1 at nanomolar concentration and even more at micromolar concentration caused the further increase of ERK1/2 activity in PC12 cells that was significant. The activation of Akt and ERK1/2 was shown to take place downstream of Trk receptor tyrosine kinase [20].

7. GM1 ganglioside normalizes the rate of respiration of PC12 cells and of mitochondria isolated from rat brain, which decreased as a result of prooxidant action

It was shown by us [23] that preincubation of PC12 cells with GM1 ganglioside prevents to a large extent the decrease of the rate of basal and uncoupled respiration, caused by the action of hydrogen peroxide. Besides, GM1 and GD1a gangliosides were shown to normalize the respiratory rates of mitochondria isolated from rat brain and exposed to prooxidant—tret-butyl hydroperoxide (tBHP). GM1 was shown to decrease the ratio of pro-apoptotic to antiapoptotic proteins Bax/Bcl-xL in control PC12 cells (from 1.0 to 0.79 ± 0.08, $p < 0.05$). As activation of ERK1/2 and Akt may lead to inactivation of proapoptotic protein Bad, which reacts with antiapoptotic proteins like Bcl-2 and Bcl-xL [24], the protective effect of GM1 and its ability to stabilize mitochondria may be a result of activation of these protein kinases [20].

GM1 and GD1a gangliosides were found to normalize the respiration rate of isolated rat brain mitochondria diminished as a result of their exposure to tBHP. It is of interest that the protective effect of these gangliosides was abolished in the presence of the inhibitor of Trk receptor tyrosine kinase K252a. In brain mitochondria, various protein kinases are present, including Trk receptor tyrosine kinase A and B [25, 26]. It was shown [26] that the protective effect of nerve growth factor on isolated brain mitochondria is abolished in the presence of the inhibitor of this protein kinase—K252a. It may be suggested that the protective effect of GM1 and other gangliosides may be to a certain extent due to their action on mitochondrial signaling pathways.

8. GM1 and GD1a gangliosides prevent TLR4 translocation into lipid rafts and protect PC12 cells from the toxic bacterial lipopolysaccharide action

Interesting data were obtained by us together with the group led by Dr. R.G. Parnova from our Institute studying the mechanism of protective effect of gangliosides against the toxic action of bacterial liposaccharide (LPS) which is the main bacterial toxin, initiating meningo-encephalites in humans. GM1 and GD1a gangliosides were shown to cause the pronounced increase of viability of PC12 exposed to toxic concentrations of LPS. Their protective effect against LPS action on PC12 cells was found not to depend on modulation of Trk receptor tyrosine kinase, and it was not revealed if PC12 cells were exposed to GM1 and GD1a in nanomolar concentration [27]. The protective effect of these gangliosides was similar to the protective effect of methyl-beta-cyclodextrin, which is known to destroy lipid rafts in cell membranes. It is known that LPS recognition and receptor complex formation occur in lipid rafts, and gangliosides play a key role in their formation and maintenance. Using subcellular fractionation, in combination with immunoblotting, and antibodies to TLR4 and flotilin (a marker of lipid rafts), it was shown that pretreatment of PC12 cells with GM1 ganglioside completely eliminated the effect of LPS on translocation of TLR4 into lipid rafts that is necessary to induce the toxic effect of LPS on the cells.

Most probably it can be explained by the fact that ganglioside incorporation in cell plasma membranes changes raft composition and properties. The results obtained suggest that ganglioside-induced prevention of TLR4 translocation into lipid rafts is a mechanism of protection against LPS action in various cells. Thus, the mechanism of protection by gangliosides against the toxic action of LPS appears to be quite different from the mechanism of their protection against the toxic effect of long incubation in serum-free medium and against glutamate, amyloid beta-peptide, or hydrogen peroxide action on the nerve cells which is mediated by activation of Trk receptor tyrosine kinase.

9. Conclusion

The considerable differences in brain lipid fatty acid composition were revealed between stenothermal cold-water and warm-water teleost fishes. The differences in fatty acid composition of brain gangliosides (which are characteristic components of neuronal plasma membranes including synaptic membranes) were found to be more pronounced between these two groups of teleosts than the differences in brain phospholipid fatty acid composition. The increase of relative content of monoenoic and polyenoic fatty acids (especially of 22:1, 24:1, and 22:6 ω3) in the brain lipids of teleost species living in cold water appears to be an important part of natural adaptations of these fishes to the conditions of their environment. Such changes in composition during adaptation to living in cold water (or at great water depth) provide the preservation of liquid-crystalline state of cell membranes and the maintenance of optimal cell membrane fluidity and microheterogeneity for the function of enzymes, receptors, and other proteins. In cartilaginous and ganoid fishes, the adaptation to living in cold water appears to be reached mainly by the increase of monoenoic fatty acid content in brain gangliosides, as polyenoic fatty acids were revealed (in low content) only in brain gangliosides of the few species of these fishes.

The correlation between the composition and structure of brain ganglioside carbohydrate component and the water temperature at which the fishes are living was not revealed. The dendrogram, constructed based on results of the cluster analysis of the data on composition of carbohydrate component of brain gangliosides of the ectothermic vertebrates, corresponds appreciably to the tree of classical taxonomy of vertebrates. The species of cartilaginous and ganoid fishes form separate clusters and are sisterly branches for each other. Species of these fishes do not form sisterly groups with any representatives of teleosts, amphibians, or reptilians. All teleost fishes are in ranges of the single separate branch on the dendrogram. The studied reptilians form a separate branch on the dendrogram.

Among the higher vertebrates, only reptilians are submitted on the cladogram. But much lower content of polysialogangliosides and much higher content of monosialogangliosides as compared to lower vertebrate brain are characteristic not only for reptilian brain but for avian and mammalian brain as well. As is known, there were cardinal modifications in the organization of animals and their separate organs at entrance of vertebrates on dry land, which are a typical example of an aromorphosis [8]. Carbohydrate components of glycolipids and glycoproteins play an essential role in cell differentiation and intercellular interaction, including cell recognition and adhesion, especially in development of organs and tissues. Modifications of the brain gangliosides composition and structure, associated with appearance of the higher vertebrates, may be suggested to provide, apparently, together with many other biochemical modifications, the molecular bases of the aromorphosis at transferring of vertebrates to the terrestrial lifestyle and amniotic development of embryos that was accompanied by differentiation and complication of brain functions.

The administration of exogenous GM1 or other main brain gangliosides (GD1a, GD1b, GT1b) to animals with the damaged brain protects brain neurons from death and improves the functional state of the animals. Various main brain gangliosides have the similar effects as typical adaptogens. The increase by GM1 of the viability of neurons and of cells of neuronal lines treated by glutamate or left in media devoid of neurotrophic factors was shown to depend on activation of Trk receptor tyrosine kinase. According to our data, GM1 ganglioside also protects neurons and PC12 cells against toxic action of amyloid peptides and hydrogen peroxide. This effect is also mediated by Trk receptor tyrosine kinase. It was shown that GM1 and other main brain gangliosides exert the protective effect on nerve cells not only at micromolar concentration but at nanomolar concentration as well, which appears to be their physiological concentration in cerebrospinal fluid and intercellular brain space. GM1 ganglioside normalizes the rate of respiration of PC12 cells and of mitochondria isolated from rat brain, decreases the proapoptotic to antiapoptotic protein ratio Bax/Bcl-xL, and activates ERK1/2 and Akt downstream of Trk receptor tyrosine kinase. Another mechanism of protective effect of GM1 and GD1a gangliosides was revealed studying their ability to increase the viability of PC12 cells exposed to bacterial LPS. In this case, the protection was not mediated by Trk receptor tyrosine kinase. The results obtained suggest that ganglioside-induced prevention of TLR4 translocation into lipid rafts is a mechanism of protection of cells against the effect of LPS, as LPS interaction with its receptor in lipid rafts is necessary to induce its toxicity. The understanding of the mechanism of the protective effect of gangliosides on nerve cells is of importance for the increase of the efficiency of clinical trials of these lipids as drugs for the treatment of ischemic, neurodegenerative, and other diseases concerned with central nervous system damage.

Acknowledgements

This work was supported by the state assignment of FASO of Russia ("The mechanisms of development of neuropsychic, metabolic, and hormonal dysfunctions in the nervous and endocrine diseases and the approaches for their correction").

Author details

Natalia F. Avrova[1*] and Yulia A. Vlasova[2]

*Address all correspondence to: avrova@iephb.ru

1 I.M. Sechenov Institute of Evolutionary Physiology and Biochemistry of the Russian Academy of Sciences, Russian Federation

2 Mechnikov North-West State Medical University, Saint-Petersburg, Russian Federation

References

[1] Kreps E, Smirnov A, Chirkovskaya E, Patrikeeva M, Krasilnikova V. Phospholipids in the nervous system of molluscs. Journal of Neurochemistry. 1968;**15**:285-291

[2] Kreps EM, Avrova N, Chebotareva M, Chirkovskaya E, Levitina M, Pomazanskaya L, Pravdina N. Some aspects of comparative biochemistry of brain lipids in teleost and elasmobranch fish. Comparative Biochemistry and Physiology. 1975;**52B**(2):293-299

[3] Kreps EM. Brain lipids of elasmobranches (an assay on comparative neurobiology). Comparative Biochemistry and Physiology. 1981;**68B**(3):363-367

[4] Avrova NF. Brain ganglioside patterns of vertebrate. Journal of Neurochemistry. 1971;**18**:667-674

[5] Avrova NF. Evolutionary approach to the analysis of structure and function of gangliosides. Neurochemical Research. 1985;**10**(11):1547-1554

[6] Avrova NF, Irwin L. New approaches to analysis of variations in brain ganglioside composition associated with idioadaptations and aromorphosis in vertebrates (in Russian). Journal of Evolutionary Biochemistry and Physiology. 2002;**38**(6):543-551

[7] Avrova NF. The comparative study of nerve tissue lipids in the works of Academician E.M. Kreps (In Russian). Journal of Evolutionary Biochemistry and Physiology. 1986; **22**(4):333-342

[8] Severtsov AS. Directedness of Evolution (in Russian). Moscow: MGU; 1990. 272 p

[9] Patterson C. Interrelationships of holosteans. In: Greenwood PH, Miles RS, Patterson C, editors. Interrelationships of Fishes. London: Academic Press; 1973. pp. 233-305

[10] Popov LS, Antonov AS. Nucleotide makeup of the DNA in representatives of certain orders of fishes (in Russian). Nauchnye Dokl Vyss Shkoly Biologicheskie Nauki 1974;**6**:33-39

[11] Zakharova IO, Avrova NF. The effect of cold stress on ganglioside fatty acid composition and ganglioside-bound sialic acid content of rat brain subcellular fractions. Journal of Thermal Biology. 2001;**26**:215-222

[12] Ferrari G, Anderson BL, Stephens RM, Kaplan DR, Greene LA. Prevention of apoptotic neuronal death by GM1 ganglioside. Involvement of Trk neurotrophin receptors. Journal of Biological Chemistry. 1995;**270**:3074-3080

[13] Bachis A, Rabin SJ, Del Fiacco M, Mocchetti I. Gangliosides prevent excitotoxicity through activation of TrkB receptor. Neurotoxicity Research. 2002;**4**:225-234

[14] Sokolova TV, Rychkova MP, Avrova NF. Protective effect of GM1 ganglioside against toxic action of glutamate on cerebellar granule cells (in Russian). Journal of Evolutionary Biochemistry and Physiology. 2014;**50**(5):399-401

[15] Sokolova TV, Zakharova IO, Furaev VV, Rychkova MP, Avrova NF. Neuroprotective effect of ganglioside GM1 on the cytotoxic action of hydrogen peroxide and amyloid beta-peptide in PC12 cells. Neurochemical Research. 2007;**32**(8):1302-1313

[16] Kreutz F, Frozza RL, Breier AC, de Oliveira VA, Horn AP, Pettenuzzo LF, Netto CA, Salbego CG, Trindade VM. Amyloid-β induced toxicity involves ganglioside expression and is sensitive to GM1 neuroprotective action. Neurochemistry International. 2011;**59**(5):648-655

[17] Kreutz F, Scherer EB, Ferreira AG, Pereira CL, Santana F, de Souza W, Salbego CG, Trindade VM. Alterations on Na+, K+-ATPase and acetylcholinesterase activities induced by amyloid-β peptide in rat brain and GM1 ganglioside neuroprotective action. Neurochemical Research. 2013;**38**(11):2342-2350

[18] Blennow K, Davidsson P, Wallin A, Fredman P, Gattfries CG, Mansson JE, Svennerholm L. Differences in cerebrospinal fluid gangliosides between "probable" Alzheimer disease and normal aging. Aging. 1992;**4**:301-306

[19] Avrova NF, Victorov IV, Tyurin VA, Zakharova IO, Sokolova TV, Andreeva NA, Stelmaschuk EV, Tyurina YY, Gonchar VS. Inhibition of glutamate-induced intensification of free radical reactions by gangliosides: Possible role in their protective effect in rat cerebellar granule cells and brain synaptosomes. Neurochemical Research. 1998;**23**:945-952

[20] Zakharova IO, Sokolova TV, Vlasova YA, Furaev VV, Rychkova MP, Avrova NF. GM1 ganglioside activates ERK1/2 and Akt downstream of Trk tyrosine kinase and protects PC12 cells against hydrogen peroxide toxicity. Neurochemical Research. 2014;**39**(11): 2262-2275

[21] Duchemin AM, Ren Q, Neff NH, Hadjiconstantinou M. GM1-induced activation of phosphatidylinositol 3-kinase: Involvement of Trk receptors. Journal of Neurochemistry. 2008;**104**:1466-1477

[22] Choi JS, Kim JA, Joo CK. Activation of MAPK and CREB by GM1 induces survival of RGCs in the retina with axotomized nerve. Investigative Ophthalmology & Visual Science. 2003;**44**(4):1747-1752

[23] Brailovskaya IV, Sokolova TV, Kobylyanskii AG, Avrova NF. Effect of ganglioside GM1 on mitochondrial respiration and viability of PC12 cells under oxidative stress conditions (in Russian). Journal of Evolutionary Biochemistry and Physiology. 2014;**50**(2):155-157

[24] Vauzour D, Vafeiadou K, Rice-Evans C, Williams RJ, Spencer JP. Activation of pro-survival Akt and ERK 1/2 signalling pathways underlie the anti-apoptotic effects of flavanones in cortical neurons. Journal of Neurochemistry. 2007;**103**:1355-1367

[25] Wiedemann FR, Siemen D, Mawrin C, Horn TF, Dietzmann K. The neurotrophin receptor TrkB is colocalized to mitochondrial membranes. The International Journal of Biochemistry & Cell Biology. 2006;**38**:610-620

[26] Carito V, Pingitore A, Cione E, Perrotta I, Mancuso D, Russo A, Genchi G, Caroleo MC. Localization of nerve growth factor (NGF) receptors in the mitochondrial compartment: Characterization and putative role. Biochimica et Biophysica Acta. 2012;**1820**(2):96-103

[27] Nikolaeva S, Bayunova L, Sokolova T, Vlasova Y, Bachteeva V, Avrova N, Parnova R. GM1 and GD1a gangliosides modulate toxic and inflammatory effects of *E. coli* lipopolysaccharide by preventing TLR4 translocation into lipid rafts Biochemica et Biophysica Acta. Molecular and Cell Biology of Lipids. 2015;**1851**(3):239-247

Chapter 3

Adaptations and Disturbances of Physiological Functions in Extreme Hyperbaric Environments

Olga S. Alekseeva, Alexander I. Krivchenko
and Ivan T. Demchenko

Additional information is available at the end of the chapter

http://dx.doi.org/10.5772/intechopen.73649

Abstract

Academician E.M. Kreps founded the Laboratory of Hyperbaric Physiology in 1960. Heads of the Laboratory were G.L. Zaltsman (1960–1972), A.I. Selivra (1972–1975), I.A. Aleksandrov (1975–1982) and I.T. Demchenko (1983–2009). In 2009, the Laboratory was merged with the Laboratory of Respiratory Physiology (A.I. Krivchenko). For more than five decades, Hyperbaric Laboratory has conducted basic and applied researches dealt with CNS oxygen toxicity, the high pressure nervous syndrome and nitrogen narcosis. Main achievements of basic researches are as follows: identified key mechanisms of adaptive responses of CNS and cardiorespiratory systems to breathing gas mixtures at high pressure, neurophysiological mechanisms of CNS oxygen toxicity and high pressure nervous syndrome, and pathogenesis of nitrogen narcosis. Main achievements of the translation of hyperbaric researches are as follows: new technology for 1000 m dive of animals (monkeys) using the gas mixture (He-N_2-O_2), new compression and decompression profiles for free escape of monkey from a depth of 700 m, use preconditional hypoxia and hyperthermia for the protection of nitrogen narcosis. Currently, main researches are focusing on the evaluation of molecular and cellular mechanisms of biological responses to extreme hyperbaric environments.

Keywords: hyperbaric oxygen, high pressure nervous syndrome, nitrogen narcosis, CNS oxygen toxicity, reactive oxygen and nitrogen species, oxidative stress, hyperoxic vasoconstriction, hyperoxic baroreflex

1. Introduction

Hyperbaric physiology researches have been conducted at the Institute of Evolutionary Physiology and Biochemistry of the Russian Academy of Sciences (IEPhB) during last 60 years and can be divided into three periods. The initial stage of hyperbaric studies is directly linked to academician Leon A. Orbeli (founder of IEPhB), who had great interest in scientific and practical aspects of the diving. Since 1935, he headed the commission for a coordination of technical projects aimed on the development of new technology for underwater diving in the USSR (former Soviet Union). After the World War II, these works were intensified markedly in order to improve the technology for raising the sunken ships, cleaning rivers, lakes and bays, and building new bridges. At that time, the breathing oxygen-helium mixture was introduced in deep sea diving. It was necessary to examine the biological effects of the helium and to develop the decompression tables for saturation diving. In 1956, academician Leon Orbeli founded the Institute of Evolutionary Physiology and organized a small group to study the biological effects of inert gases under pressure. According to Orbeli's conception, hyperbaric biological research in addition to practical aspects will help to substantiate the role of extreme environmental factors in the evolution of physiological functions. In 1960, new director of the IEPhB academician Evgeny M. Kreps organized the Laboratory for Environmental Physiology and invited Henry L. Saltzman, a former naval doctor, to conduct physiological studies in hyperbaria. For next 8 years, the world standard results were obtained regarding the physiological responses to oxygen, helium, and nitrogen under high pressure [1].

By the mid 1960s, a new trend in the hyperbaric researches has been appeared. George F. Bond (US NAVY officer) offered a new technology that consists of the pressure chamber is installed on the board of ship, and divers living in hyperbaric conditions for several weeks are delivered to the sea bottom by special bell with subsequent return to the pressurized chamber. Academician E.M. Kreps has initiated a discussion of novel diving technology at Russian Academy of Sciences, which resulted in a new building, the installation of hyperbaric facilities and substantially expanded stuff. Since then, the second phase of hyperbaric researches began in the Institute. Animals (guinea pigs, rats and rabbits) were exposed to normoxic helium-oxygen gas up to 41 ATA for 2–3 weeks following physiological and morphological examinations. Many experts were involved in these studies, and the results of 10 years of experimental work were summarized in the book [2]. From 1972 to 1983, experimental studies were focused on the biological effects of indifferent gases, toxic effects of hyperbaric oxygen and morphological changes and functional disorders after prolonged hyperbaric exposures. At the same time, academician E. Kreps initiated the construction of new hyperbaric facilities. In 1983, the Laboratory was headed by Professor Ivan T. Demchenko. One year later, a new hyperbaric complex "KIZH-100" has been put into operation. KIZH-100 was specially designed for hyperbaric studies on animals when breathing gas mixtures are at pressures up to 101 ATA (atmospheres absolute). The number of staff of Hyperbaric Laboratory has been increased to more than 70 specialists in 1985. It has been purchased and installed world-class equipment for physiological and biochemical studies on animals and for the efficient operation of chambers. The main subject for hyperbaric studies were the monkeys (Macaca irus). In the short time, new approaches and methods were developed for experiments on monkeys with chronically implanted electrodes and probes. The continuous exposure of monkeys in helium-nitrogen-oxygen mixtures under high pressure was up to 5 weeks. The focus of the research was to explore the possibility of 1000 m diving with a protection of the high pressure nervous syndrome. The results were published by Russian Physiological Journal [3] and later in the Proceeding of III International Meeting "High Pressure

Biology," Durham, USA, 1993 [4] and in the V International Meeting "High Pressure Biology and Medicine," St. Petersburg, Russia, 1997 [5]. The experimental results obtained in KIZH-100 remain unique in the international research practice. Unfortunately, the economic situation in Russia since the 1990s did not allow continuing saturation diving studies.

The third period of hyperbaric researches in IEPhB started since 1991. Because of limited funding, the chamber saturation diving studies were discontinued. Main efforts have been directed to study cellular and molecular mechanisms of action of helium, nitrogen and oxygen under pressure. Valery B. Kostkin evaluated the molecular effects of hyperbaric environmental factors [6]. Alexander N. Vetosh and Olga S. Alekseeva studied the biological effects of hyperbaric nitrogen at normoxic and hypoxic conditions [7–10]. Hyperbaric studies in the Institute have been supported by the Foreign Members of the Russian Academy of Sciences, Professor Peter B. Bennett and Professor Claude A. Piantadosi (Duke University, Durham, USA). Important support for hyperbaric research was provided by its former employees: D.N. Atochin (Harvard University, Boston, USA) and D.R. Gutsaeva, (Medical College of Georgia, Augusta, USA). Long-time cooperation with American colleagues has been productive in studying molecular mechanisms of hyperbaric oxygen toxicity.

2. CNS O_2 toxicity

Over the past 50 years, Hyperbaric Physiology Lab has carried out basic researches to explore mechanisms by which hyperbaric oxygen (HBO_2) elicits CNS O_2 toxicity. The objectives of these studies were: (1) to determine the temporal and spatial profiles of reactive oxygen and nitrogen species (RONS) in the brain during the development of HBO_2-induced seizures; (2) to evaluate the physiological and pathological responses to HBO_2 related to oxygen seizures development and (3) to identify the crucial large-scale sites for CNS O_2 toxicity. Achievements of these studies were published (see References) and are briefly described here as follows:

(1) Concise accomplishments of studies. We have confirmed and expanded the concept that primary HBO_2-derived originators initiated toxic effects on CNS are reactive oxygen and nitrogen species (RONS). We measured RONS ($^{\bullet}O_2^-$, OH^-, H_2O_2, $NO^{\bullet}$, and $ONOO^-$) in the brain during HBO_2 exposures at 5–6 ATA and found their excessive production and progressive accumulation, as a function of the inspired oxygen partial pressures and the time of exposure [11–13]. We clarified that one root component of RONS is superoxide anion excessively produced in HBO_2 as a by-product of the nonspecific transfer of electrons to O_2 by either mitochondrial electron transport proteins or by non-mitochondrial enzymes. Brain H_2O_2 and OH^- levels increased in HBO_2 as a result of excessive $^{\bullet}O_2^-$ production and superoxide dismutase (SOD) and catalase activity [13]. Nitric oxide ($NO^{\bullet}$) is the other root component of RONS excessively produced in HBO_2 by L-arginine biotransformation involving endothelial and neuronal NOS, as demonstrated on knock-out mice [14]. HBO_2 stimulates the production of $^{\bullet}O_2^-$ and $NO^{\bullet}$ leading to an increase in peroxynitrite ($ONOO^-$) formation prior to seizures. Neuronal NOS-derived $NO^{\bullet}$ generates the bulk of brain $ONOO^-$, which mediates neurotoxic effects of HBO_2 [13]. We also found different rates of $^{\bullet}O_2^-$ - and $NO^{\bullet}$ accumulation in the brain and showed that an emergence of $^{\bullet}O_2^-/NO^{\bullet}$ imbalance is crucial for oxygen seizures development [15]. $NO^{\bullet}$ is exquisitely sensitive to

inactivation by $^{\bullet}O_2^{-}$, and extracellular superoxide dismutase (SOD3) regulates $NO^{\bullet}$ bioavailability [15]. Thus, excessive formation and accumulation of RONS in the brain are inevitable in HBO_2 and can be considered as a trigger for the development of CNS O_2 toxicity.

(2) Concise accomplishments of studies. Physiological and pathological responses to HBO_2 have been determined as follows. CNS responds to HBO_2 by a progressive disruption of the brain's normal electrical activity finally manifested by generalizing EEG spikes [1, 16, 17]. The appearance of EEG spikes (seizures), as a sign of CNS O_2 toxicity, always followed by excessive RONS accumulation in the brain [11, 13]. Any interventions limiting RONS production or their scavenging prevented EEG seizures. For example, systemic nonselective NOS inhibition with L-NAME prevented O_2 seizures [16, 19], and the source of $NO^{\bullet}$-mediated HBO_2 seizures is neuronal NOS as demonstrated experiments on gene knockout ($nNOS^{-/-}$, $eNOS^{-/-}$ and $iNOS^{-/-}$) mice [13]. Transgenic mice with SOD3 overexpression showed higher susceptibility to HBO_2 than wild type demonstrating important role of $^{\bullet}O_2^{-}/NO^{\bullet}$ balance for CNS O_2 toxicity [15]. Thus, EEG spiking activity is mediated by RONS accumulation in the brain, and the hypersynchronization of neuronal firing signifies the brain overexcitation and impairment of CNS functions. CNS-derived somatic (motor) responses in dogs, cats, rabbits, rats and mice to HBO_2 are manifested by the progressive behavioral and motor disturbances such as intensive grooming, local jerks, "wet-dog" shakes, fast-running and finally tonic and clonic convulsions with the loss of consciousness. These disturbances correlated with specific EEG patterns and final convulsions always followed EEG spikes with short delay [1, 20]. Rats treated with myorelaxants (tubocurarine or pancuronium bromide) did not exhibit any motor responses in HBO_2 at 5 or 6 ATA but paroxysmal EEG activity has been manifested [16]. We think that any local or generalized motor jerks or convulsions (seizures) in HBO_2 are secondary to CNS dysfunction and not direct effects of hyperbaric oxygen on skeletal muscle. Supporting findings of this point is that systemic nonselective NOS inhibition with L-NAME prevented both EEG seizures and motor convulsions in HBO_2 [16, 19].

The lungs' responses to hyperoxia are O_2 pressure dependent. In $HBO_2 \leq 2.5$ ATA, motor convulsions or EEG seizures have never been observed at least for 16 h exposures [21]. Because the entire surface of the lung is directly exposed to the hyperoxic environment for many hours, the inflammatory responses are developed slowly with destruction of the alveolar-capillary barrier, edema, impaired gas exchange, respiratory failure and death. However, the lungs respond to O_2 at 5 or 6 ATA by acute damage [22]. Pulmonary damage, characterized by transpulmonary leakage of protein and focally distributed intra-alveolar hemorrhage, developed rapidly, and key factors for lung injury were left ventricular function impairment and abruptly elevated pulmonary venous pressure leading to the cardiogenic lung injury [22]. Systemic NOS inhibition protected against HBO_2-induced pulmonary damage in $eNOS^{-/-}$ and $iNOS^{-/-}$ mutants compared to that seen in wild-type (WT) mice, but $nNOS^{-/-}$ mutants were relatively protected [14]. Collectively, these findings demonstrate that neuronal NOS (nNOS) play a prevalent role in the development of HBO_2 pulmonary toxicity [19].

Responses of autonomic nervous system (ANS) to HBO_2 are O_2 partial pressure and time exposure dependent. When HBO_2 does not exceed 2.5 ATA (clinical HBO settings), the parasympathetic (vagal) activity dominates as indicated, a decrease in heart rate, cardiac output and respiration. In this case, ANS controls cardiovascular and pulmonary functions reflexively mainly through baroreflex activation triggered by a rise in arterial pressure due to hyperoxic vasoconstriction [23]. Resulting afferent discharges from the arterial baroreceptors evoke central responses that suppress efferent sympathetic activity and augment parasympathetic outflow providing short-term adaptation to hyperoxic environments. HBO_2 exposures exceeding 3 ATA also initially exhibit prevalent vagal tone but then autonomic imbalance appears in favor of sympathetic activation and parasympathetic withdrawal. Progressively increased sympathetic outflow affects cardiac function and pulmonary hemodynamic, leading to lung injury [16].

Cardiovascular system responds to HBO_2 by systemic vasoconstriction, bradycardia, cardiac output reduction and redistribution of organ blood flow. Moderate HBO_2 (<2.5 ATA) induces cerebral vasoconstriction that is associated with increased $^{\bullet}O_2^-$ production and a decrease in $NO^{\bullet}$ availability around the vessel's smoth muscle. Hyperoxic vasoconstriction is attenuated in $SOD3^{+/+}$ mutant or eNOS-/- mice demonstrating critical role of $^{\bullet}O_2^-/NO^{\bullet}$ balance in cerebrovascular responses to HBO_2 [15]. Extreme HBO_2 (>3 ATA) induces biphasic CBF response: transient vasoconstriction followed by hyperemia [16–18]. HBO_2-stimulated $NO^{\bullet}$ production increased CBF and oxygen delivery prior to the appearance of EEG spikes. Transgenic $SOD3^{+/+}$ mice are more sensitive, while $eNOS^{-/-}$ mice are more resistant to HBO_2-induced seizures. HBO_2 seizures are associated with an increase in cerebral blood flow (CBF) that hastens the onset of convulsions through delivery of a toxic oxygen dose. Endothelial-derived $NO^{\bullet}$ has a principal contribution to the development of hyperemia preceding O_2 seizures.

(3) Concise accomplishments of studies. Hyperbaric studies have identified neuronal discharges related to HBO_2 seizures both in CNS and periphery. Earlier neurophysiological studies of CNS O_2 toxicity have shown that the development of oxygen EEG seizures is dynamic process that comprises three distinct stages. The first stage characterized by a formation of single unstable foci of neuronal excitation in subcortex centers (reticular formation, thalamus and hypothalamus), following the appearance of stable and multiple foci in mesodiencephalic parts of the brain during the preconvulsive second EEG stage [1]. Finally, in the third convulsive stage, the process terminates in a synchronization of paroxysmal EEG activity in all parts of the brain [1]. Stages in EEG seizures correlate with a balance between excitatory and inhibitory neurotransmission. Our work demonstrated that the brain excitability in HBO_2 is occurred first in subcortical structures after a significant reduction in extracellular GABA content and a minor increase in glutamate [25]. The Glu/GABA imbalance appears to be the critical trigger for the hypersynchronization of neuronal firing manifested as EEG spiking activity. CNS O_2 toxicity is linked also with the autonomic nervous system. Neuronal endings (receptors) stimulated or inhibited by an alteration in arterial PO_2, intravascular pressure and RONS in visceral tissues initiate three types of reflexes such as chemoreflex, baroreflex and cardiac sympathetic afferent reflex, by which ANS modulates brain excitation. We showed that afferent signaling from aortal and carotid baroreceptors normally restrain brain excitability, but in HBO_2 5 or 6 ATA, baroreflex is impaired and seizure latency shortened [24]. Cerebrovascular responses to HBO_2 affected by RONS are also critical contributors to CNS O_2 toxicity. Hyperoxic vasoconstriction decreases CBF and delays seizures, but cerebral hyperemia accelerates seizure development through the alterations of oxygen content delivery.

Thus, our studies outlined here dealt with temporal and spatial RONS accumulation in the brain during extreme hyperoxic exposures, RONS-related physiological and pathological responses in HBO_2 and CNS O_2 toxicity initiation and progression. However, an overall view of CNS O_2 toxicity should also be comprised primary targets affected by RONS, in particular, their location, molecular structures and mechanisms of RONS-target interactions. All of these issues are remained still obscure and are the object of future studies.

3. High pressure nervous syndrome

High pressure nervous syndrome (HPNS) is a neurological disorder occurs when man dives below 150 m using helium-containing breathing gas. The severity of HPNS depends on the rate of descent, the depth and the percentage of helium. First noted in the 1960s, HPNS was referred as "helium tremor." Helium tremor was reported by G.L. Zaltsman in his human studies since 1961 [26]; however, this experimental fact was not available in English-language publications until 1967 [27]. At the same time, P. Bennett investigated helium tremor and widely described its patterns in 1965 [28]. The term "high pressure nervous syndrome" was introduced by Brauer in 1968 to describe the combined symptoms of tremor, electroencephalography (EEG) changes, and somnolence that appeared during a 362 m chamber dive [29, 30]. Main symptoms of HPNS in humans are dizziness, nausea, vomiting, postural and intention tremors, fatigue and somnolence, myoclonic jerking, stomach cramps, decrements in intellectual and psychomotor performance, poor sleep with nightmares, and increased slow wave and decreased fast wave activity on electroencephalogram [31]. In animal study, HPNS is manifested by tremor, myoclonic jerks, convulsions and specific patterns on EEG including spiking activity [31].

In Hyperbaric Physiology Lab, HPNS was evaluated in experiments on monkeys exposed to heliox at 101 ATA. Polarographic measurements demonstrated physiological levels of brain PO_2 at 101 ATA, when oxygen pressure in inspired gaseous mixture was 0.35 ATA, but decreased in normoxic heliox mixture [4, 5, 32]. Monkeys have shown the signs of HPNS (tremor upper limbs, specific EEG pattern) at 20–25 ATA but these symptoms were delayed by adding nitrogen (7 or 10%) to heliox. Using neuropharmacological approaches, Alexandr Sledkov has shown that HPNS manifestation is a result of an increase in brain excitability and the threshold for excitability, in various brain structures was different [33]. In rabbits, the lowest threshold response to increased helium pressure (about 15 ATA) had the limbic system and especially, the hippocampus [33]. This neuropharmacological study also showed that the adrenergic system does not involve in the HPNS pathogenesis, and its pharmacological activation does not lead to an alteration in functional activity of the subcortex. Activation of dopaminergic system prevents the development of HPNS manifestations. The serotonergic system may play a role in the mechanisms of HPNS manifestations, but the hypothesis of its compliance with the so-called "serotonin syndrome" has not been confirmed. Activation of the cholinergic system in hyperbaric conditions is very dangerous because of the sharp drop in the sensitivity thresholds and seizures. Apparently, HPNS implementation that takes place chiefly through N-cholinergic has confirmed the protective effect of N-cholinolytics, whose mechanism of action is based on reducing the level of hippocampal excitability, when the suppression of muscarinic receptor type comes with potentiation effect of HPNS. Activating the GABAergic system plays a protective role in HPNS. The mechanism of this protective effect of brake amino acids, possibly nonspecific and acts by enhancing inhibitory and excitatory inhibition processes in the CNS [33].

4. Hyperbaric nitrogen narcosis

Nitrogen narcosis is a condition that occurs in divers when breathing compressed air. Behnke et al. were the first to prove that the nitrogen in compressed air is responsible for signs and symptoms of narcosis, characterized as "euphoria, retardation of the higher mental processes and impaired neuromuscular coordination" [31]. When divers go below the depths of approximately 30 m, an increase in the partial pressure of nitrogen alters mental state similar to alcohol intoxication. This discovery stimulated a research of biological effects of hyperbaric nitrogen in Military Medical Academy in Leningrad, Russia [1]. In the Laboratory of Hyperbaric Physiology, nitrogen narcosis studies have been carried out since 1960s on three lines.

The first line of research concerned the electrophysiological analysis of CNS responses to hyperbaric nitrogen breathing. Zaltzman et al. investigated nitrogen, argon and helium narcosis in animals (dogs, rabbits and mice) by multichannel EEG recording. The results of EEG analysis in rabbits have shown that air pressure at 5 ATA suppressed the alpha wave but the beta activity increased. At pressure of 8 ATA, EEG exhibited slow theta activity, and an exposure to 12 ATA led to pronounced suppression of EEG activity [1]. The first EEG changes in the dogs were observed under the pressure of argon-5 atm, nitrogen-10 atm and helium-15 atm. The EEG patterns in hyperbaric argon and nitrogen were similar. The progression of EEG changes had three stages such as the depression of cortex activity, the domination of theta rhythm in the brain subcortex with the transition to delta activity, and the generalization of delta activity. At 35–40 atm of nitrogen, EEG generalization was unstable. A peculiar feature of the hyperbaric helium was that the theta rhythm in the brain stem structures developed and generalized without any preliminary suppression of the activity of cortex against a background of increased activity in the structures of the striatal system [1].

The second line of research concerned the determination of physiological and biochemical correlates of nitrogen narcosis. Alexander Vetosh has found behavior correlates in the progression of nitrogen narcosis [7]. He determined the patterns of motor activity and posture reflexes in mammals exposed to hyperbaric nitrogen and developed quantified scale of nitrogen narcosis levels. It was established experimentally that mammals can maintain vital functions while breathing oxygen-nitrogen mixtures at density up to 151 g/l. This is 117 times greater than the density of air. Continuing this line of research, O. Alekseeva et al. found the biochemical markers of the nitrogen narcosis. They also reported an increase in heat shock proteins of HSP-70 family in nitrogen narcosis stage [8, 9].

The third line of research concerned the mechanisms of nitrogen narcosis. According to the literature, the primary molecular mechanism of nitrogen narcosis is based on the ultrastructural changes in biological membranes of neurons in the brain due to excess dissolved nitrogen in their lipids [31]. A new suggestion about the mechanism of nitrogen narcosis was offered by A. Vetosh [7] and expanded by O. Alekseeva [8]. They showed a correlation between nitric oxide ($NO^{\bullet}$) formation, generation of heat shock proteins and the progression of nitrogen narcosis. They suggested that RONS formed in hyperbaric nitrogen are implicated in narcosis through the alteration in cellular function leading to motor, emotional and cognitive symptoms of nitrogen narcosis. Studies showed that L-NAME, nonselective inhibitor of NOS, significantly delayed the nitrogen narcosis symptoms [7]. Hypoxic preconditioning before nitrogen diving mobilized HSP-70 family of proteins in blood and brain, and delayed the signs of nitrogen narcosis by 67% [8, 9]. Thus, the problem of biological action of hyperbaric nitrogen has a long history, and we think that the abovementioned experimental data contribute to the understanding of the pathogenesis of nitrogen narcosis as well as the creation of technological and pharmacological methods of its correction.

Extreme hyperbaric environments perturb various cellular processes at the molecular level due to the effects of pressure per se, gas partial pressure alone, through an intensive production of reactive oxygen and nitrogen species (ROS/RNS), which can incorporate in redox signaling pathways stimulating adaptive physiological responses or damaging cellular machinery. Altered pressure environments are routinely encountered in hyperbaric medicine (hyperbaric oxygen therapy) and diving (hyperbaric gases), and next basic research will focus on the obtaining data for better understanding of these potential applications.

Acknowledgements

This work was supported by State budget funding according to the assignment by the Russian Federal Agency for Scientific Organizations (FASO Russia) for Section 6 ("Physiological mechanisms of adaptation of man and animals to extreme and periodically changing geo-geliophysical and meteorological factors").

Author details

Olga Sergeevna Alekseeva[1], Alexander Ivanovich Krivchenko[1] and
Ivan Timofeyevich Demchenko[1,2]*

*Address all correspondence to: itdemch@mail.ru

1 Sechenov Institute of Evolutionary Physiology and Biochemistry of the Russian Academy of Sciences, St. Petersburg, Russia

2 Duke University Center for Hyperbaric Medicine and Environmental Physiology, Durham, NC, USA

References

[1] Zaltsman GL, editor. Hyperbaric Epilepsy and Narcosis (Neurophysiological Studies). Leningrad: Nauka; 1968. 267 p

[2] Kreps EM, Zaltsman GL, editors. The organism in conditions of prolonged hyperbaric conditions: Research experience in experimental models: Collected articles. Leningrad: Nauka; 1977. 192 p

[3] Demchenko IT. Hyperbaric physiology studies. Russian Physiological Journal. 1992;**78**(3): 46-57

[4] Bennett PB, Marquis RE, editors. Basic and Applied High Pressure Biology. NY, USA: University of Rochester Press; 1994. 441 p

[5] Bennett PB, Demchenko IT, Marquis RE, editors. High Pressure Biology and Medicine. NY, USA: University of Rochester Press; 1998. 415 p

[6] Zabelinskii SA, Chebotareva MA, Kostkin VB, Demchenko IT, Krivchenko AI, Khozhai LI, Otellin VA. Phospholipids composition of cell nuclei of rat's brain in dynamics of normal ontogenesis and after hypoxia. Doklady Biological Sciences. 2005;**400**(6):10-12

[7] Vetosh AN. Nitrogen Influence on Biological Systems. St. Petersburg, Russia: Nauka; 2003. 231 p

[8] Alekseeva OS, Vetosh AN, Kostkin VB, Korzhevskiy DE, Otellin VA. Heat shock proteins in brain neurons and hypoxic preconditioning. Doklady Biological Sciences. 2009;**425**(1):98-100

[9] Alekseeva OS, Vetosh AN, Kostkin VB, Korzhevskii DE. Influence of quercetin on the progress of nitrogen narcosis and accumulation of heat shock proteins in cells of the rat cerebral cortex. Doklady Biological Sciences. 2010;**430**(1):11-13

[10] Alekseeva OS, Vetosh AN, Kostkin VB, Korzhevskii DE. Preadaptation to nitrogen anesthesia and impairment of rats brain cortex structure during hypoxia. Journal of Evolutionary Biochemistry and Physiology. 2010;**46**(4):374-378

[11] Demchenko IT, Boso AE, Whorton AR, Piantadosi CA. Nitric oxide production is enhanced in rat brain before oxygen-induced convulsions. Brain Research. 2001;**917**(2):253-261

[12] Gutsaeva DR, Suliman HB, Carraway MS, Demchenko IT, Piantadosi CA. Oxygen-induced mitochondrial biogenesis in the rat hippocampus. Neuroscience. 2006;**137**(2):493-504

[13] Demchenko IT, Atochin DN, Boso AE, Astern J, Huang PL, Piantadosi CA. Oxygen seizure latency and peroxynitrite formation in mice lacking neuronal or endothelial nitric oxide synthases. Neuroscience Letters. 2003;**344**(1):53-64

[14] Demchenko IT, Atochin DN, Gutsaeva DR, Godfrey RR, Huang PL, Piantadosi CA, Allen BW. Contributions of nitric oxide synthase isoforms to pulmonary oxygen toxicity, local vs. mediated effects. American Journal of Physiology-Lung Cellular and Molecular Physiology. 2008;**294**(5):984-990

[15] Demchenko IT, Oury TD, Crapo JD, Piantadosi CA. Regulation of the brain's vascular responses to oxygen. Circulation Research. 2002;**91**:1031-1037

[16] Demchenko IT, Boso AE, O'Neill TJ, Bennett PB, Piantadosi CA. Nitric oxide and cerebral blood flow responses to hyperbaric oxygen. Journal of Applied Physiology. 2000;**88**(4): 1381-1389

[17] Demchenko IT, Bosso AE, Zhiliaev SY, Moskvin AN, Gutsaeva DR, Atochin DN, Bennett PB, Piantadossi CA. Involvement of nitrogen oxide in the cerebral vasoconstriction

during respiration with high pressure oxygen. Rossiĭskii Fiziologicheskiĭ Zhurnal Imeni I. M. Sechenova Rossiĭskaia Akademiia Nauk. 2000;**86**(12):1594-1603

[18] Demchenko IT, Luchakov YI, Moskvin AN, Gutsaeva DR, Allen BW, Thalmann ED, Piantadosi CA. Cerebral blood flow and brain oxygenation in rats breathing oxygen under pressure. Journal of Cerebral Blood Flow & Metabolism. 2005;**25**(10):1288-1300

[19] Allen BW, Demchenko IT, Piantadosi CA. Two faces of nitric oxide: implications for cellular mechanisms of oxygen toxicity. Journal of Applied Physiology. 2009;**106**(2):662-667

[20] Demchenko IT, Boso AE, Natoli MJ, Doar PO, O'Neill TJ, Bennett PB, Piantadosi CA. Measurement of cerebral blood flow in rats and mice by hydrogen clearance during hyperbaric oxygen exposure. Undersea and Hyperbaric Medicine. 1998;**25**(3):147-152

[21] Demchenko IT, Welty-Wolf KE, Allen BW, Piantadosi CA. Similar but not the same: normobaric and hyperbaric pulmonary oxygen toxicity, the role of nitric oxide. American Journal of Physiology – Lung Cellular and Molecular Physiology. 2007;**293**(1):229-238

[22] Demchenko IT, Zhilyaev SY, Moskvin AN, Piantadosi CA, Allen BW. Autonomic activation links CNS oxygen toxicity to acute cardiogenic pulmonary injury. American Journal of Physiology – Lung Cellular and Molecular Physiology. 2011;**300**(1):102-111

[23] Demchenko IT, Moskvin AN, Krivchenko AI, Piantadosi CA, Allen BW. Nitric oxide-mediated central sympathetic excitation promotes CNS and pulmonary O_2 toxicity. Journal of Applied Physiology. 2012;**112**(11):1814-1823

[24] Demchenko IT, Zhilyaev SY, Moskvin AN, Krivchenko AI, Piantadosi CA, Allen BW. Baroreflex-mediated cardiovascular responses to hyperbaric oxygen. Journal of Applied Physiology. 2013;**115**(6):819-828

[25] Demchenko IT, Piantadosi CA. Nitric oxide amplifies the excitatory to inhibitory neurotransmitter imbalance accelerating oxygen seizures. Undersea and Hyperbaric Medicine. 2006;**33**(3):169-174

[26] Zaltsman GL. Physiological Foundations of Human Presence in a High-pressure Gaseous Medium. Moscow: Medgiz; 1961. 188 p

[27] Zaltsman GL. Psychological Principles of a Sojourn of a Human in Conditions of Raised Pressure of the Gaseous Medium (in Russian, 1961). English translation, Foreign Technology Division. AD655 360. Ohio: Wright Patterson Air Force Base; 1967

[28] Bennett PB. Psychometric Impairment in Men Breathing Oxygen-Helium at Increased Pressures. Royal Navy Personnel Research Committee, Underwater Physiology Subcommittee Report No. 251. London: Medical Research Councill, 1965. 14 p

[29] Brauer RW, Dimov S, Fructus X, Fructus P, Gosset A, Naquet R. Syndrome neurologique et electrographique des hautes pressions. Revue Neurologique (Paris). 1968;**121**(3):264-265

[30] Brauer RW. Seeking man's depth level. Ocean Industry (London). 1968;**3**:28-33

[31] Bennett PB, Rostain JC. The high pressure nervous syndrome. In: Brubakk AO, Neuman TS, editors. Bennett and Elliott's Physiology and Medicine of Diving. 5th Rev ed. United States: Saunders; 2003. pp. 323-357

[32] Atochin DN, Zhilyaev SY, Krivchenko AI. Brain oxygenation in gunea pigs breathing gaseous mixture at high density. Russian Physiological Journal. 1992;**78**:53-64

[33] Sledkov AY. Neuropharmacological bases of development and prevent HPNS and nitro gen narcosis. St.-Petersburg, Russia: Ivanov & Leschinsky Press; 1995. 226 p

Chapter 4

Evolution of Thalamic Sensory Centers in Amniotes: Phylogeny and Functional Adaptation

Margarita G. Belekhova and Natalia B. Kenigfest

Additional information is available at the end of the chapter

http://dx.doi.org/10.5772/intechopen.73650

Abstract

This chapter is a continuation of our previous study of the forebrain evolution in vertebrates using some new tests allowing evolutionary transformations to be revealed. As such tests, we chose the expression of calcium-binding proteins as neuronal functional markers and the metabolic activity of cytochrome oxidase, characterizing the level of neuronal activity. Here, we report the results of our study of the thalamic visual and auditory centers in reptiles (turtles, *Emys orbicularis* and *Testudo horsfieldii*) and birds (pigeon, *Columba livia*) with a special focus on differences in their parallel visual thalamofugal and tectofugal channels and auditory lemniscal and extralemniscal channels. A comparison with data obtained in other Sauropsida amniotes was drawn to elucidate the role of phylogenetic and functionally adaptive factors determining variable distribution of calcium-binding proteins and metabolic activity, as well as to identify evolutionary conservative and plastic traits in the organization of these thalamic sensory centers.

Keywords: visual system, auditory system, Ca-binding proteins, metabolic activity, Sauropsida amniotes, evolution

1. Introduction

Since the creation of Charles Darwin's theory of evolution [1], most studies in the field of evolutionary neuroscience were focused mainly on the phylogenetic continuity in the evolution of the central nervous system in vertebrates. A central problem of comparative neurobiology was a search for homologous brain structures in different taxa of vertebrates through the identification of common ancestral (plesiomorphic) and acquired (apomorphic) traits (see for references [2–4]). At the same time remarkable diversity of brain structures in every vertebrate divergent lineage is a result of two evolutionary pathways – phylogenetic history and adaptive specialization (apomorphosis and idioadaptation according to Severtsov [5]). The second pathway is a key in the origin of homoplasy (parallel and convergent evolution). Two non-antagonistic but rather complementary approaches – historical (phylogeny) and causal (evolving adaptive mechanisms) Dobzhansky considered necessary for the evolutionary synthesis [6]. A combination the embryogenetic, morphological, histochemical and functional approaches as well as the introduction of modern molecular and genetic methods led to a partial or even total revision of some classical views of brain evolution. One of the most crucial achievements was the revision of the old concept of the homology between the basal ganglia and isocortex in Sauropsida amniotes and mammals. According to the new concept, a great part of the avian and reptilian telencephalon, previously considered as a homolog of the striatum, has a pallial origin and is homologous to the mammalian cortex. Like in mammals, the basal ganglia (striatopallidum) occupy only the ventral part of the telencephalic hemisphere. On this basis, the nomenclature of telencephalic structures in birds has been modified [7–10]. Birds were thus rehabilitated as possessors of the highly developed pallium as compared to the telencephalic cortex in mammals. Behavioral studies conducted in various species of birds and reptiles also led to the reevaluation of their cognitive capabilities, in some avian species as compared to those in primates [9, 11–13]. However, the issues of what parts of the mammalian cortex are homologous to the avian telencephalic pallial parts targeted by the thalamic relay nuclei still remain a matter of indefatigable debate.

While the homology of the thalamofugal (geniculocortical) pathway in amniotes is now generally accepted, two alternative hypotheses have been advanced regarding the homology of the thalamopallial tectofugal and auditory pathways. According to the "neocortex hypothesis, " the thalamic projection fields in the pallium of birds and reptiles are homologous to the mammalian isocortex (a dorsal pallium derivative) [7, 8, 10, 14, 15]. The "claustroamygdalar hypothesis" draws a homology between them and a part of the claustroamygdalar complex (a ventral/lateral pallium derivative) [4, 16, 17]. Respectively, thalamic projection nuclei in reptiles and birds are comparable either with dorsothalamic relay nuclei in mammals ("neocortex hypothesis") [8, 10, 14, 15] or a part of the thalamic complex of intralaminar and posterior nuclei ("claustroamygdalar hypothesis") [4, 16, 18]. There is no final solution for this problem.

2. Results and their evolutionary implications

Phylogenetic transformations in the sensory thalamo-telencephalic systems were considered by the classics of comparative neurology as critical for understanding the forebrain evolution. In the laboratory of A.I. Karamian, the visual, auditory and somatosensory systems were investigated for many years (1958–1989) in the wide range representatives of different vertebrate classes. It was established that these systems consist of parallel pathways, having different morphological and functional characteristics, and different rates of phylogenetic development (see Refs. [19–22]).

We are carrying out comparative studies of the visual and auditory systems in amniotes, and birds (Archosauria), descending from a common ancestor and thus, having a key significance for understanding the forebrain evolution. As new complementary tests, characterizing the organization of the visual and auditory centers, we used: (1) immunohistochemical analysis of expression of parvalbumin (PV) and calbindin (CB), calcium-binding proteins serving as functionally selective neuronal markers and (2) histochemical evaluation of cytochrome oxidase (CO) metabolic activity reflecting the level of neuronal functional activity. Expression of calcium-binding proteins was studied using the standard procedure of immunohistochemistry on free-floating 40 μm sections. Monoclonal mouse anti-PV (Sigma, USA) diluted 1:1000 and polyclonal rabbit anti-CB (Swant, Switzerland) diluted 1:5000 were used. Cytochrome oxidase activity was revealed on free-floating 40 μm sections according to the convenient histochemical method using cytochrome c from bovine heart, type III (Sigma, USA) as well. Sections were observed and analyzed using the microscope Zeiss Axio Imager A1 (Zeiss, Germany). Images were taken from representative sections with the digital camera mounted on the microscope. Digital images were created using Adobe Photoshop 7.0 (Adobe Systems Incorporated, USA) and assembled into montages. General adjustments of color, contrast, and brightness were made.

This chapter offers a brief comparative survey of our previously published and recently obtained results in the thalamic visual and auditory centers in turtles and pigeons, as well as their analysis in the light of the relevant literature data and present knowledge in this field. We set ourselves the task of elucidating: (1) to what extent the patterns of PV and CB immunoreactivity coincide in homologous centers of reptiles and birds; (2) whether the expression of PV and CB correlates with CO activity; and (3) whether these data can shed light on the role of the phylogenetic and functional (adaptive) factors in determining the PV and CB specificity of the sensory centers.

2.1. Visual system

Across all sauropsids (nonarchosaurian reptiles, Archosauria: birds and crocodiles), the visual system includes two main pathways projecting to the telencephalon, tecto- and thalamofugal, that have different properties. Within the tectofugal pathway in reptiles and birds, projections of retinal ganglion cells successively relay in the optic tectum, thalamic nucleus rotundus (Rot), which further projects to the visual dorsolateral region of the anterior dorsal ventricular ridge (ADVRdl) in reptiles and to the entopallium (Ent) in birds. Within the thalamofugal pathway, retinal ganglion cells directly project to the thalamic relay nucleus geniculatus lateralis, pars dorsalis (GLd), which then projects, in reptiles, to the dorsolateral cortex and, in birds, to the hyperpallial Wulst. Homology between these systems in reptiles and birds is generally accepted. They are comparable, according to "neocortex hypothesis, " to the mammalian thalamo (nucleus lateralis posterior+pulvinar)-extrastriate and geniculo-striate systems, respectively [3, 8–10].

Our results show that the tectofugal thalamic center Rot has a higher metabolic (CO) activity both in turtles (**Figure 1a-c**) and pigeons (**Figure 1d**, **f**) as compared to the thalamofugal center GLd. This correlates with the leading role of the tectofugal (collothalamic and extralemniscal) visual system in visual behavior [3]. Differences in the level of CO activity between the Rot and GLd are less significant in pigeons, probably due to a more highly developed thalamofugal visual system in birds.

As for the CaBPr immunoreactivity, the Rot in reptiles and birds differs by the distribution, ratio of PV- and CB-ir neurons, and intensity of their labeling. In the turtle Rot, strongly labeled

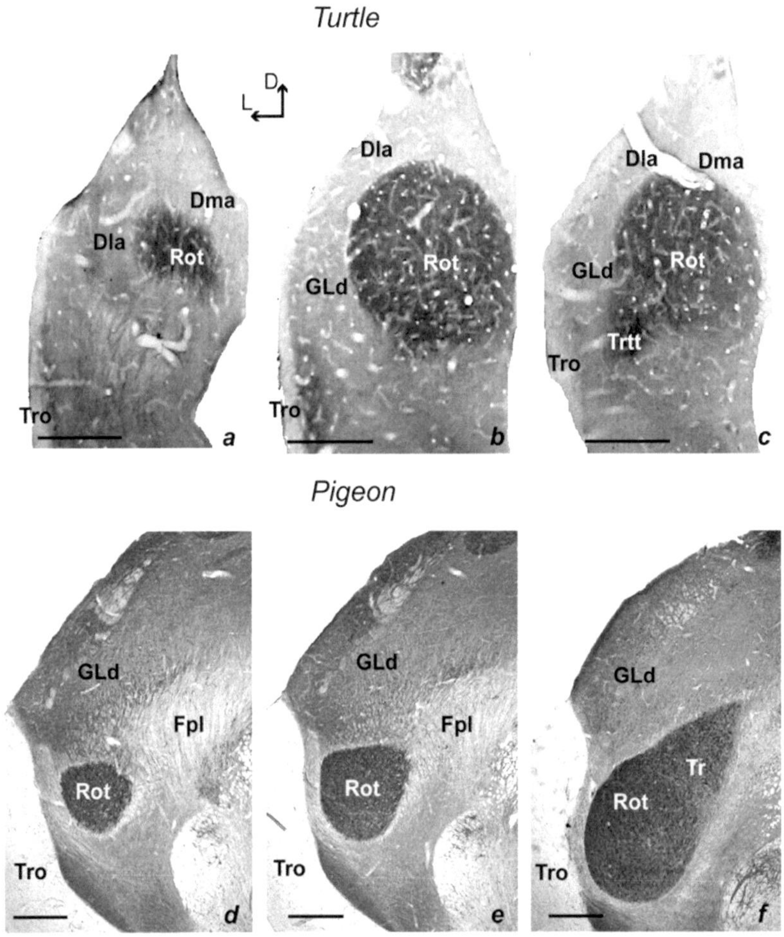

Figure 1. The activity of cytochrome oxidase in visual centers of turtles and pigeons. Rostrocaudal transverse unilateral sections of the thalamus in turtle (a-c) and pigeon (d-f). Note a high CO activity in the rot and a weaker CO activity in the GLd, both in turtle and pigeon. CO—Cytochrome oxidase; Dla—N. Dorsolateralis anterior; Dma—N. Dorsomedialis anterior; Fpl fasciculus prosencephali lateralis; GLd—*N. geniculatus* Lateralis, pars dorsalis; rot—*N. rotundus*; Tr—*N. triangularis*; Tro—Tractus opticus; Trtt—Tractus tectothalamicus. D—Dorsal; and L—lateral sides. Dorsal and lateral sides are the same here and in other figures. Scale bar: 500 μm.

CB-ir cells prevailed (**Figures 2b** and **3a**), whereas PV-ir cells were less numerous (**Figures 2b** and **3b**). On the contrary, in the pigeon Rot, strongly labeled PV-ir neurons were prevailing, whereas CB-ir cells exhibited a restricted distribution pattern and mainly weak labeling (**Figures 2d** and **3c**). In the triangular part of the Rot (Tr), strongly labeled PV- and CB-ir cells were observed to overlap (**Figures 2d** and **3c**, **d**). According to multiple studies in other reptilian and avian species, a great interspecies variability was found in the number of PV- and CB-ir neurons, ranging from the mixed content of both types to the existence of only one of them (see for Refs. [23, 24]). At the same time, both in turtles and pigeons, the Rot has an abundant PV innervation (**Figure 3d** along with a high CO activity (**Figure 1a–f**). In birds, the tectorotundal pathway contains multiple parallel channels, deriving from different types of

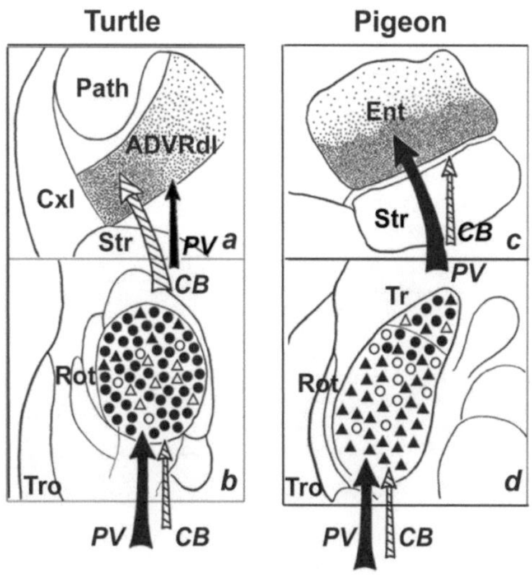

Figure 2. Different specificity to parvalbumin and calbindin of rotundo-telencephalic pathways in turtle (a, b) and pigeon (c, d). Schematic drawings of unilateral transverse sections of the brain at the level of the rot (b, d) and telencephalic areas (a, c), receiving projections from the rot. Circles indicate CB-ir neurons, triangles indicate PV-ir neurons (black are for strongly, white are for weakly labeled cells), and dots indicate immunoreactive terminals. Black arrows mark PV-ir input and striped arrows mark CB-ir input. ADVRdl—Dorsolateral anterior dorsal ventricular ridge; CB—Calbindin; cxl—Cortex lateralis; Ent—Entopallium; path—Pallial thickening; PV—Parvalbumin; rot—*N. rotundus*; Str—Striatum; Tr—*N. triangularis*; And Tro—Tractus opticus.

tectal neurons [25, 26] and processing different aspects of visual information [27, 28]. We found that tectorotundal projection neurons in birds and reptiles expressed PV and CB [29, 30]. Thus, a heterogeneous distribution of PV and CB immunoreactivity in the avian and reptilian Rot may relate to different chemospecificity of parallel tectorotundal channels.

Both in the turtle GLd (**Figure 3a, b** and the largest GLd subnuclei (DLAmc, DLL) of pigeons (**Figure 3e, f**), strongly labeled CB-ir neurons prevailed with PV-ir cells being less numerous. The other avian GLd subnuclei were found to contain cells immunoreactive either to both proteins or only to PV, as in the LdOPT [24]. At the same time, both in turtles and pigeons, CO activity in the GLd was lower than in the Rot with an exception for the LdOPT, where it was very high. Similar to the Rot, there is a great interspecies variability in the patterns of PV and CB immunoreactivity in the GLd of reptiles and birds (see for references [24]).

In turtles, rotundal PV- and CB-ir neurons project to the ADVRdl (**Figure 2a**, **b**); in pigeons, rotundal neuronal projections terminate in the Ent (**Figure 2c, d**). Geniculate neurons immunoreactive to these proteins project to the dorsolateral cortex in turtles (see details in [23, 31, 32]) and to the Wulst in birds (see Ref. [24]). The density of telencephalic innervation (immunoreactive dotted neuropil) positively correlates with the number of corresponding immunoreactive cells in the projection thalamic nuclei [23, 31, 32].

The prevalence of PV expression in the Rot and CB expression in the GLd in the zebra finch [33, 34] allowed concluding that in birds, the tectofugal system (Rot-Ent) is PV-specific, while the thalamofugal system (GLd-Wulst) is CB-specific. By contrast, in the comparable visual

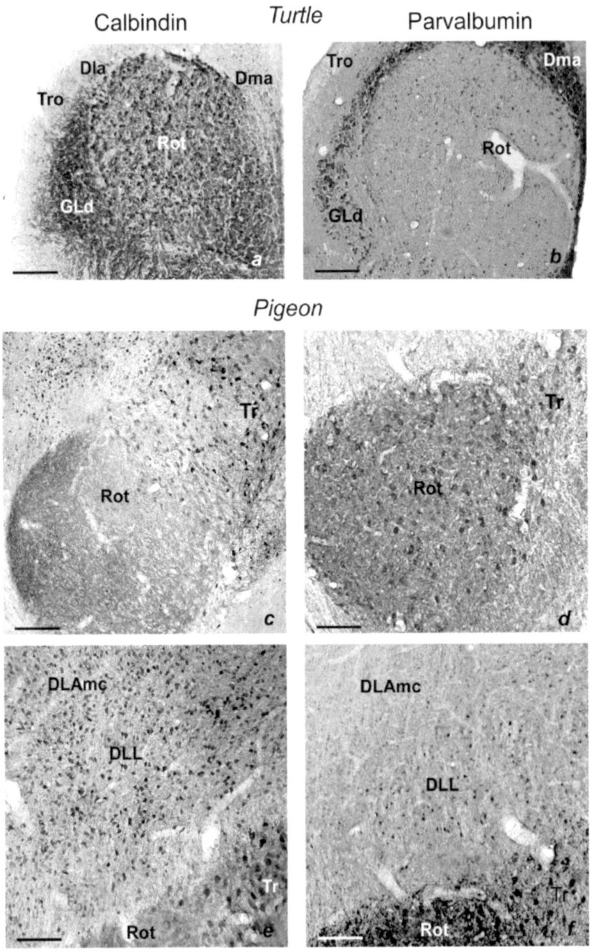

Figure 3. Patterns of parvalbumin and calbindin immunoreactivity in the rot and GLd in turtle and pigeon. Microphotographs of unilateral transverse sections of the thalamus in turtle (a, b) and pigeon (c-f)). Note the prevalence of CB-ir (a) and lesser number of PV-ir (b) cells in the turtle rot in contrast to the prevalence of PV-ir (d) and lesser number of CB-ir cells (c) in the pigeon rot. Both in turtle GLd (a, b) and pigeon GLd—DLAmc, DLL (e, f), prevalence of CB-ir cells (a, e) and lesser number of PV-ir cells (b, f). CB—Calbindin; Dla—N. Dorsolateralis anterior; DLAmc—N. Dorsolateralis anterior magnocellularis; DLL—N. Dorsolateralis anterior lateralis, pars lateralis; Dma—N. Dorsomedialis anterior; GLd—*N. geniculatus* Lateralis, pars dorsalis; PV—Parvalbumin; rot—*N. rotundus*; Tr—*N. triangularis*; And Tro—Tractus opticus. Scale bar: 100 μm.

pathways of mammals [34], the distribution of PV and CB is quite opposite: the extrageniculocortical system (LP/Pulv-extrastriate cortex) is CB-specific. However, the data obtained in the zebra finch cannot be transposed to all avian species because there is a great interspecies variability in the pattern of CaBPr immunoreactivity in the centers of the tecto- and thalamofugal pathways (see Ref. [24]). Similar variability exists in the reptilian thalamic centers of the tecto- and thalamofugal systems, being mainly CB-specific in both cases [23, 24, 31, 32, 35]. Thus, the examples of both similarity and dissimilarity in PV and CB immunoreactivity can be found in homologous visual thalamic centers of reptiles and birds. Here, we disregard the expression of other CaBPr, although, for example, calretinin has been demonstrated in the visual and auditory thalamic centers in reptiles and birds [23, 32, 34–36].

A study of CaBPr in the thalamus of higher mammals (primates) allowed E. Jones [37] to put forward a hypothesis that PV prevails in the phylogenetically younger, highly specialized lemniscal (core) centers, whereas CB is predominant in the phylogenetically older, less specialized structures (matrix), including the extralemniscal regions (belt/shell) of the sensory nuclei. These findings and the data on CaBPr in the brain structures of nonprimate mammals [37–39], Sauropsida amniotes and anamniote vertebrates (see [24]), led to a conclusion that distribution of different types of PV- and CB-expressing neurons in brain structures depends on the level of phylogenetic development. However, a high variability in the neuronal PV and CB immunoreactivity in the lemniscal parts of the homologous thalamic sensory nuclei in amniotes, including nonprimate mammals, revealed numerous exceptions of the Jones' concept. Altogether, they have led us to conclude that at every stage of phylogenetic history, the specificity to different CaBPr types depends on the functional factor (see discussion in [24]).

2.2. Auditory system

The auditory system in all amniotes contains two parallel pathways such as lemniscal and extralemniscal. Both of them derive from the mesencephalic auditory center, but from its different regions: the lemniscal stems from the core region, while the extralemniscal—from the peripheral belt region. The lemniscal pathway projects to the core (Red+Revm) of the thalamic auditory center nucleus reuniens (Re) in reptiles and to the core (nCe Ov) of the nucleus ovoidalis (Ov) in birds. The extralemniscal pathway projects to the peripheral regions of these nuclei, respectively, to the Revl in reptiles and the Ovl and Ovm in birds. Both pathways have different morphological, neurochemical, and functional characteristics and different targets in the auditory telencephalic regions: the lemniscal—in the core (central area of the ADVRvm in reptiles, L2 in birds), whereas the extralemniscal—in the belt (peripheral area of ADVRvm in reptiles, L1, L3, CMM in birds) [40–48].

The distribution of PV and CB immunoreactivity as well as CO activity was different in the central and peripheral regions of the thalamic auditory centers in turtles and pigeons, reflecting their core-belt organization. In turtles, the core region (Red+Revm) contains both CB- and PV-ir cells as well as a neuropil with prevailing CB immunoreactivity (**Figure 4b, c**), and exhibits high CO activity (**Figure 4a**). The belt region (Revl) is distinguished by a weak immunoreactivity to both proteins and a low CO activity (**Figure 4a–c**). The prevalence of CB-ir cells in Red+Revm positively correlates with a high density of CB-ir neuropil in its projection telencephalic field (ADVRvm) that decreased at the border with the ADVRm (**Figure 4e**). PV immunoreactivity of neuropil in the ADVRvm was far less dense, while CO activity was rather high, but only outside of neuronal clusters (**Figure 4d**).

Pigeons have a more distinct core-belt organization of the thalamic auditory center Ov as compared to the turtle Re. The Ov core region (nCe Ov), like the turtle Red+Revm, contains both CB- and PV-ir cells and neuropil, but with prevailing PV immunoreactivity. The density of dotted PV-ir neuropil and the degree of cell labeling therein were greater (**Figures 5c, d** and **6a**)

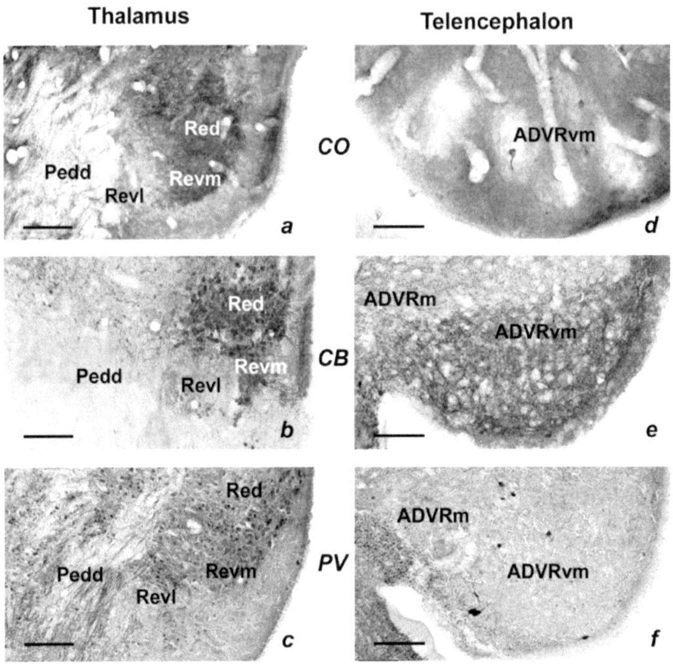

Figure 4. Distribution of calbindin and parvalbumin immunoreactivity and CO activity in the thalamic (re) and telencephalic (ADVRvm) auditory centers in turtle. Transverse unilateral sections at the levels of re (a-c) and ADVRvm (d-f)). A, d—CO activity, b, e—CB, c, f—PV immunoreactivity. Note the highest level of both CB and PV immunoreactivity and CO activity in the red+Revm (core of the re) in contrast to the Revl (belt of the re). Compare the ADVRvm both strongly CB-ir (e) and moderately CO-active (d) terminal neuropil with weakly PV-ir neuropil (f). White areas in d—CO-negative cell clusters. ADVR s– Anterior dorsal ventricular ridge; ADVRm—Medial part of ADVR; ADVRvm—Ventromedial part of ADVR; CB—Calbindin; CO—Cytochrome oxidase; Pedd—Pedunculus dorsalis; PV—Parvalbumin; re—N. Reuniens; red —Re dorsalis; Revl—Re ventrolateralis; and Revm—Re ventromedialis. Scale bar: 100 μm.

than in CB-ir neuropil and its cells (**Figures 5e, f** and **6b**). A high CO activity of neuropil and its cells clearly distinguished the nCe Ov from the peripheral nuclei Ovl and Ovm, where this activity was absent (**Figure 5a, b**).

Like in mammals, the ratio and distribution of CB- and PV-ir neurons in the lemniscal (core) regions of the Re and Ov significantly differ not only across different Sauropsida taxa but also in different species within the same taxonomic group [33, 36, 37, 48, 49]. These interspecies differences relate to peculiarities in the morphofunctional organization of the lemniscal centers in different species, specifically with different localization of brain stem auditory input projections, encoding information about different parameters of sound signaling. Overall, the variability in CaBPr expression in the lemniscal (core) centers is determined by specific mechanisms for processing each auditory modality. Thus, the phenotypic diversity in the CaBPr expression in lemniscal auditory centers may be considered as a result of the complicated interplay between phylogenetic history and ecology-dependent functional specialization with the leading role of the functionally adaptive factor.

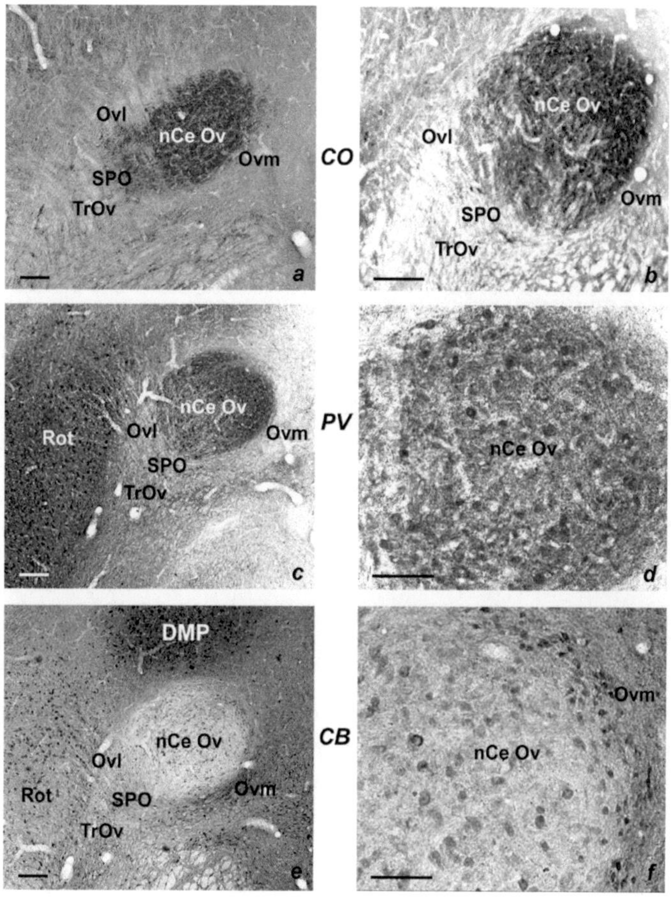

Figure 5. Distribution of calbindin and parvalbumin immunoreactivity and CO activity in the thalamic auditory center (Ov) of the pigeon. Transverse unilateral thalamic sections at the level of the Ov. (a, b)—CO activity located only in the nCe Ov (core). (c, d)—Strong PV immunoreactivity located only in the nCe Ov. (e, f)—CB immunoreactivity located in the nCe Ov, Ovl, and Ovm. Note strongly stained PV-ir and weakly stained CB-ir cells in the nCe Ov (core) in contrast to strongly CB-ir cells in Ovl and Ovm (belt), which are morphologically different from CB-ir cells in the nCe Ov. Note also a high level of CB immunoreactivity in the DMP (e). CB—Calbindin; CO—Cytochrome oxidase; DMP—N. Dorsomedialis posterior; nCe OV—*N. centralis* Ov; Ov—N. Ovoidalis; Ovl—Ov lateralis; Ovm—OV medialis; PV—Parvalbumin; rot—*N. rotundus*; SPO — Nucleus semilunaris parovoidalis; and TrOv—Tractus ovoidalis. Scale bars: 100 μm (a, c, e) and 50 μm (b, d, f).

In the extralemniscal peripheral Ov regions (Ovl and Ovm) of pigeons, a distinct monospecificity to CB was revealed. These nuclei contained only CB-ir cells and dense CB-ir neuropil (**Figures 5e, f** and **6b**), being completely devoid of PV immunoreactivity (**Figures 5c** and **6a**). At the same time, CB-ir cells were more densely packed, strongly labeled, and exhibited a different morphological type as compared to CB-ir cells in the nCe Ov (**Figures 5f** and **6b**). This feature is typical for the belt Ov regions in all the studied avian species and for some belt nuclei in the mammalian auditory thalamic center (nucleus geniculatus medialis) (see for Ref. [48]). Such a strong similarity indicates a high evolutionary conservatism of the extralemniscal auditory thalamic center. It is determined by the fact that peripheral parts of the auditory

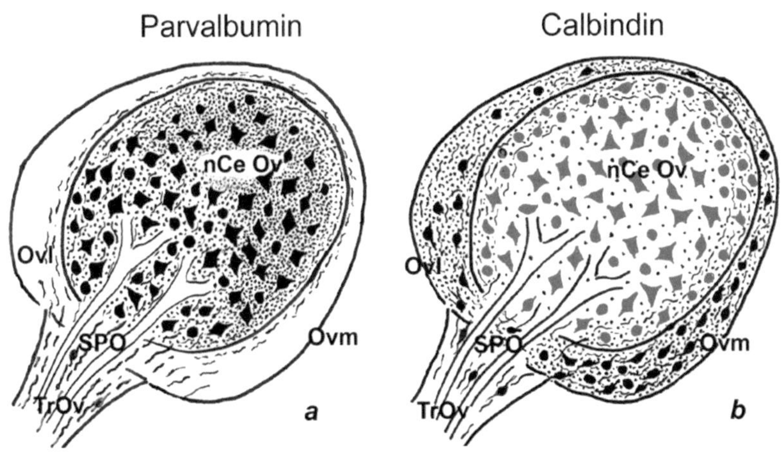

Figure 6. Core-belt organization of the pigeon nucleus ovoidalis. Schematic drawings of transverse sections of the Ov. The core nCe Ov contains both PV-ir (a) and CB-ir (b) cells and dotted neuropil. Note: Strong labeling of PV-ir and weaker labeling of CB-ir cells, high density of PV-ir neuropil, and low density of CB-ir neuropil. Belt Ovl, Ovm, and SPO contain only strongly labeled CB-ir cells and neuropil (b) and devoid of PV-ir cells (a). CB—Calbindin; nCe OV—*N. centralis* Ov; Ov—N. Ovoidalis; Ovl—Ov lateralis; Ovm—OV medialis; PV—Parvalbumin; SPO — Nucleus semilunaris parovoidalis; and TrOv—Tractus ovoidalis.

centers have multiple connections with many other nonauditory, including limbic centers, which provide the involvement of auditory information in different vital functions of the brain, responsible for feeding, reproductive, communicative, and other behaviors served species survival [43, 44, 46, 50].

3. Conclusion

In reptiles and birds (Archosauria), the patterns of calcium-binding protein (PV and CB) expression and metabolic (CO) activity have been shown to differ in distinct areas of the visual and auditory thalamic centers related to different parallel channels within the tecto- and thalamofugal visual pathways as well as the lemniscal (core) and extralemniscal (belt) auditory pathways. No unambiguous positive correlation has been found in the thalamic centers between PV immunoreactivity and high CO activity. The level of metabolic activity is likely to depend on the functional significance of the thalamic centers. The remarkable interspecies variability in PV and CB expression in homologous centers within every phylogenetic lineage appears to result from complicated interrelationships between phylogeny and epigenetic ecology-dependent functional adaptation, reflecting both conservative and plastic traits in their evolutionary development. The patterns of PV and CB immunoreactivity in the thalamic centers of the reptilian and avian visual and auditory systems provide evidence in favor of their homology with the mammalian dorsothalamic projection nuclei and, accordingly, the homology of their projection pallial areas with the mammalian isocortical sensory zones, supporting thereby the Karten's isocortical hypothesis [7, 10].

Acknowledgements

This work was supported by State budget funding according to the assignment by the Russian Federal Agency for Scientific Organizations (FASO Russia) (Neurophysiological mechanisms of functional regulation and their evolution).

Author details

Margarita G. Belekhova[1]* and Natalia B. Kenigfest[1,2]

*Address all correspondence to: belekhova@yahoo.com

1 I.M. Sechenov Institute of Evolutionary Physiology and Biochemistry, Russian Academy of Sciences, Saint-Petersburg, Russia

2 Museum National d'Histoire Naturelle USM-0501, Paris, France

References

[1] Darwin C. On the Origin of Species by Means of Natural Selection. London: Murray; 1859

[2] Northcutt RG. Understanding vertebrate brain evolution. Integrative and Comparative Biology. 2002;**42**:743-756

[3] Butler AB, Hodos W. Comparative Vertebrate Neuroanatomy. Evolution and Adaptation. 2nd ed. Hoboken, New Jersey: Wiley; 2005

[4] Striedter GF. Principles of Brain Evolution. Irvine: University of California; 2005. 357 p

[5] Severtsov AN. Main Trends in the Evolutionary Process. Morphobiological Evolutionary Theory. (Book in Russian). Biomedgiz: Leningrad; 1934

[6] Dobzhansky T. Genetics and the Origin of Species. New York: Columbia University Press; 1951

[7] Nauta WH, Karten HJ. A general profile of the vertebrate brain, with sidelights on the ancestry of cerebral cortex. In: The Neurosciences: Second Study Program. New York: Rockfeller Univ. Press; 1970. pp. 7-26

[8] Reiner A, Yamamoto K, Karten HJ. Organization and evolution of the avian brain. The Anatomical Record. 2005;**287**:1080-1102

[9] Jarvis ED, Güntürkün O, Bruce LL, Csillag A, Karten HJ, Kuenzel W, Medina L et al. Avian brains and a new understanding of vertebrate brain evolution. Nature Reviews. Neuroscience 2005;**6**:151-159

[10] Karten HJ. Vertebrate brains and evolutionary connections: On the origins of the mammalian "neocortex". Philosophical Transactions of Royal Society B. 2015;**370**. DOI: 10.1098/rstb.2015.0060

[11] Emery NJ. Cognitive ornithology: The evolution of avian intelligence. Philosophical Transactions of Royal Society B. 2006;**361**:23-43

[12] Northcutt RG. Variation in reptilian brains and cognition. Brain, Behavior and Evolution. 2013;**82**:45-54

[13] Kis A, Huber L, Wilkinson A. Social learning by imitation in a reptile (*Pogona utficeps*). Animal Cognition. 2015;**18**:325-331

[14] Butler AN, Reiner A, Karten HJ. Evolution of amniote pallium and the origins of mammalian neocortex. Annals of the New York Academy of Sciences. 2011;**1225**:14-27

[15] Dugas-Ford J, Ragsdale CW. Levels of homology and the problem of the neocortex. Annual Review of Neuroscience (Palo Alto, CA). 2015;**38**:351-368

[16] Bruce LL, Kornblum HI, Seroogy KB. Comparison of thalamic populations in mammals and birds: Expression of ErbB4 mRNA. Brain Research Bulletin. 2002;**57**:455-461

[17] Puelles L. Thoughts on the development, structure and evolution of the mammalian and avian telencephalic pallium. Philosophical Transactions of Royal Society B. 2001;**356**:1583-1598

[18] Guirado S, Real MA, Davila JC. The ascending tectofugal visual system in amniotes: New insights. Brain Research Bulletin. 2005;**66**:290-296

[19] Karamian AI. Functional Evolution of the Vertebrate Brain (Monograph in Russian). Leningrad: Nauka; 1970. 304 p

[20] Karamian AI. Evolution of the Vertebrate Telencephalon (Monograph in Russian). Nauka: Leningrad; 1976. 256 p

[21] Vesselkin NP. Evolution of Afferent Systems at Early Phylogenetic Stages in Vertebrates (Doctoral Thesis in Russian). Leningrad; 1974

[22] Belekhova MG. Thalamotelencephalic System of Reptiles (Monograph in Russian). Leningrad: Nauka; 1977. 216 p

[23] Belekhova MG, Kenigfest NB, Minakova MN; Rio J-P, Reperant J. Calcium-binding proteins in the thalamus of turtles. Analysis of the thalamic organization according to the theory of "core-matrix" and in the connection with the problem of homology of the amniote thalamic nuclei. Journal of Evolutionary Biochemistry and Physiology 2003;**39:**624–648

[24] Belekhova MG, Chudinova TV, Rio J-P, Tostivint H, Vesselkin NP, Kenigfest NB. Distribution of calcium-binding proteins in the pigeon visual thalamic centers and related pretectal and mesencephalic nuclei in pigeons. Phylogenetic and functional determinants. Brain Research. 2016;**1631**:165-193

[25] Karten HJ, Cox K, Mpodozis J. Two distinct populations of tectal neurons have unique connections within the retinotectorotundal pathway of the pigeon (*Columba livia*). The Journal of Comparative Neurology. 1997;**387**:449-465

[26] Hellmann B, Güntürkün O. Structural organization of parallel information processing within the tectofugal visual system of the pigeon. The Journal of Comparative Neurology. 2001;**429**:94-112

[27] Wang Y-C, Jiang S, Frost B. Visual processing in pigeon nucleus rotundus: Luminance, color, motion and looming subdivisions. Visual Neuroscience. 1993;**10**:21-30

[28] Laverghetta AV, Shimizu T. Visual discrimination in the pigeon (*Columba livia*): Effects of selective lesions of the nucleus rotundus. Neuroreport. 1999;**10**:981-985

[29] Chudinova TV, Kenigfest NB, Belekhova MG. Components of the pigeon tectothalamic visual pathway, revealed with aid of study of cytochrome oxidase activity and immunoreactivity to calcium-binding proteins. Journal of Evolutionary Biochemistry and Physiology. 2010;**46**:622-630

[30] Belekhova MG, Chudinova TV, Kenigfest NB. Calcium-binding proteins and cytochrome oxidase activity in the turtle optic tectum with special reference to the tectofugal visual pathway. Journal of Evolutionary Biochemistry and Physiology. 2013;**49**:519-540

[31] Belekhova MG, Kenigfest NB, Chudinova TV. Calcium-binding proteins and metabolic activity (cytochrome oxidase) in thalamotelencephalic pathways of the turtle visual system. Journal of Evolutionary Biochemistry and Physiology. 2012;**48**:322-334

[32] Kenigfest NB, Belekhova MG. Neurons of visual thalamic nuclei projecting to telencephalon express different types of calcium-binding proteins: A combined immunocytochemical and tracer study. Journal of Evolutionary Biochemistry and Physiology. 2015;**51**:505-516

[33] Heizmann CW, Braun K. Calcium binding proteins. Molecular and functional aspects. In: Anghileri LJ, editor. The Role of Calcium in Biological Systems. Boca Raton, FL: CRC Press Inc.; 1990. pp. 21-65

[34] Heyers D, Manns M, Luksch H, Gűntűrkűn O, Mouritsen H. Calcium-binding proteins label functional streams of the visual system in a songbird. Brain Research Bulletin. 2008; **75**:324-355

[35] Davila JC, Guirado S, Puelles L. Expression of calcium-binding proteins in the diencephalon of the lizard *Psammodromus algiris*. The Journal of Comparative Neurology. 2000;**427**:67-92

[36] Yan K, Tang Y-Z, Carr CE. Calcium-binding protein immunoreactiviry characterizes the auditory system of *Gekko gecko*. The Journal of Comparative Neurology. 2010;**518**:3409-3426

[37] Jones EG. Viewpoint: The core and matrix of thalamic organization. Neuroscience. 1998; **85**:331-345

[38] Diamond IT, Fitzpatrick D, Schmechel D. Calcium-binding proteins distinguish large and small cells of the ventral posterior and lateral geniculate nuclei of the prosimian galago and tree shrew (*Tupaia belangeri*). Proceedings of the National Academy of Sciences of the United States of America. 1993;**90**:1425-1429

[39] Hof PR, Glezer TT, Conde F, Flagg RA, Rubin MB, Nimchinsky EA, Vogt Weisenhorn DM. Cellular distribution of the calcium-binding proteins parvalbumin, calbindin, and calretinin in the neocortex of mammals: Phylogenetic and developmental patterns. Journal of Chemical Neuroanatomy. 1999;**16**:77-116

[40] Karten HJ. The ascending auditory pathway in the pigeon (*Columba livia*) II. Telencephalic projections of the nucleus ovoidalis thalami. Brain Research. 1968;**11**:134-163

[41] Pritz MB. Ascending connections of a thalamic auditory area in a crocodile *Caiman crocodilus*. The Journal of Comparative Neurology. 1974;**153**:199-213

[42] Pritz MB, Stritzel ME. A second auditory area in the non-cortical telencephalon of a reptile. Brain Research. 1992;**569**:146-151

[43] Durand SE, Tepper JM, Cheng MF. The shell region of the nucleus ovoidalis: A subdivision of avian auditory thalamus. The Journal of Comparative Neurology. 1992;**323**:495-518

[44] Wild JM, Karten HJ, Frost BJ. Connections of the auditory forebrain in the pigeon (*Columba livia*). The Journal of Comparative Neurology. 1993;**337**:32-62

[45] Belekhova MG, Kenigfest NB, Vesselkin NP, Rio J-P, Reperant J, Ward R. Evolutionary significance of different neurochemical organization of the internal and external regions of auditory centers in the reptilian brain: An immunocytochemical and reduced NADPH-diaphorase histochemical study in turtles. Brain Research. 2002;**925**:100-106

[46] Zeng S, Zhang X, Peng W, Zuo MX. Immunohistochemistry and neural connectivity of the Ov shell in song-bird and their evolutionary implications. The Journal of Comparative Neurology. 2004;**470**:192-209

[47] Zeng S, Li J, Zhang X. Distinction of neurochemistry between core and their shells of auditory nuclei in tetrapod species. Brain, Behav. Evol. 2007;**70**:1-20

[48] Belekhova MG, Chudinova TV, Repérant J, Ward R, Jay B, Vesselkin NP, Kenigfest NB. Core-and-belt organisation of the mesencephalic and forebrain auditory centres in turtles: Expression of calcium-binding proteins and metabolic activity. Brain Research. 2010;**13**: 84-102

[49] Chudinova TV, Belekhova MG, Tostivint E, Rio J-P, Ward R, Kenigfest NB. Differences in CB- and PV-chemospecificity in the centres of the ascending auditory pathway of turtles revealed by double immunofluorescence labeling. Brain Research. 2012;**1473**:87-103

[50] Cheng MF, Peng JP. Reciprocal call between the auditory thalamus and the hypothalamus: An antidromic study. Neuroreport. 1997;**8**:653-658

Discovery of the Phenomenon of Intracellular Development of Cardiac Stem Cell: A New Step in Understanding of Biology and Behavior of Tissue-Specific Stem Cells

Galina B. Belostotskaya, Tatyana A. Golovanova,
Irina V. Nerubatskaya and Michael M. Galagudza

Additional information is available at the end of the chapter

Abstract

In our experiments with an in vitro culture of rat cardiac cells, we identified and described for the first time the phenomenon of intracellular development of CSCs in mature CMs with formation of the "cell-in-cell structures" (CICSs). Recently, we have confirmed the reproducibility of our results and existence of this phenomenon in rats of different age groups, 1-year-old bull, adult mice and humans. Moreover, we demonstrated the 5–10 times increase in the amount of CICSs after exposure of in vitro cultures to hypoxia and acidosis, that is, these conditions stimulate intracellular development of CSCs. Our data strongly suggest that transitory amplifying cells (TACs), which release from CICSs, are present as a very rare cell population in adult and old rats. Therefore, we assume that TACs are important for renewal of myocardium during ontogenesis. TACs should be considered as the major source of cells that can reduce myocardial damage in adult mammals with various pathologies of the cardiovascular system. In conclusion, precise and exhaustive analysis of the phenomenon of intracellular development of CSCs, CICSs and TACs will pave the way for cell technologies of new generation in regenerative medicine.

Keywords: primary culture of myocardial cells, mature cardiac myocytes, resident cardiac stem cells, proliferation, differentiation, cell-in-cell structure, transitory amplifying cells

1. Introduction

The founder of the study of stem cells (SCs) in adult organism is A.A. Maximow, who in 1909 first used the term «stem cell» and proved the existence of pluripotent hematopoietic stem cell publishing a paper «The Lymphocyte as a stem cell common to different blood elements in embryonic development and during the post-fetal life of mammals» [1]. However, the ideas of our fellow-countryman A.A. Maximow, a Professor at the Saint Petersburg Imperial Military Medical Academy and Head of the Department of Histology and Embryology, were developed only in the 1960s, when Canadian scientists E. McCulloch and J. Till found hematopoietic SCs in the bone marrow [2]. The next breakthrough in this direction was made by Russian scientist A.Ya. Friedenstein, a Corresponding Member of the Academy of Medical Sciences of the U.S.S.R. [3]. He found in the bone marrow not only hematopoietic, but also stromal SCs, that can turn into cells of bone, cartilage, fibrous and fatty tissue. Going back to the ideas of A.A. Maximow, it should be emphasized that he suggested the existence of progenitor cells in certain tissues, marking them as stem mesenchymal cells. In this concept, he not only put their embryonic origin, but also their differentiation potency, ability to progressive differentiation into all kinds of connective tissue in postnatal development, such as reparative histogenesis.

Basic elements of the stem cell concept, developed in the study of the hematopoietic system, have been further extended to other rapidly renewing organs and tissues. It was found that the undifferentiated tissue-specific SCs (progenitor cells) are located in various organs and tissues, such as skin [4], cornea [5], kidney [6], liver [7], tooth [8, 9] and others, and that they are responsible for renewal of their cell population, replacing dead cells.

Although the detailed discussion of the characteristics of tissue-specific SCs is beyond the scope of this review, the brief history of neuronal SC discovery is described below as an example of long-lasting debate between "classical" views and new original ideas. Despite the fact that a paper by J. Altman, titled «Are new neurons formed in the brains of adult mammals?» [10] and published in the Science Magazine in 1962, and a few other publications by him, have proved the existence of neurogenesis in the adult mammalian brain, a concept that «the nerve cells do not regenerate» still persisted. Only twenty years later neurogenesis was again «discovered», but in the brains of birds. In the mid-1980s, Professor F. Nottebohm at Rockefeller University (U.S.A.) was able to show that in the vocal center of adult male canaries, a neurogenesis process occurs constantly in varying degrees [11]. In the late 1980s and early 1990s neuronal SCs have also been found in adult amphibians at the laboratory of our Institute of Evolutionary Physiology and Biochemistry of the U.S.S.R. Academy of Sciences, led by physiologist, endocrinologist and morphologist, a Corresponding Member of the Academy of Sciences, Andrey L. Polenov [12, 13]. In the same time period, neuronal progenitor cells were identified and studied first in the embryos [14], and then in the adult vertebrate animals [15]. These studies, along with the following works, have led to recognition of the existence of neuronal SCs, that has become not only an important fundamental discovery in stem cell biology, but also open up the prospects for their use in the treatment of peripheral nerve injury [16], spinal cord trauma [17, 18], acute cerebral injury [19], and neurodegenerative diseases, such as Alzheimer disease and multiple sclerosis [20–22].

2. Which cells are involved in the renewal and regeneration of the mammalian myocardium?

Paradoxically, mammalian heart was considered as an organ not capable of self-renewal until the beginning of the XXI century. It was widely believed that the cardiomyocytes formed in the first days after birth, persist throughout life and cannot be replaced in case of damage or destruction, and an increase in heart size from birth to adulthood is due to hypertrophy of cardiomyocytes but not to their proliferation [23].

The first signs of a possible myocardial recovery, manifested in the resumption of DNA synthesis, have been observed in the studies of Russian scientists in the late 1970s-80s [24–26], while the data on increased telomerase activity, stimulation of cyclins and cyclin-dependent kinases, and enhanced proliferation of myocytes in mammals, including humans, in the later stages of cardiac insufficiency after myocardial infarction began to appear at the turn of twentieth and twenty-first centuries [27–30].

Existence of the SCs in adult mammalian heart and their participation in remodeling of injured myocardium was revealed for the first time only in 2002 in the paper by Hierlihy et al. [31] and later papers [23, 32–35]. A number of studies performed on the culture of heart cells, yielded valuable information about types of the resident CSCs, their clonogenicity and ability to differentiate [32, 36–41]. Also, it was found that cardiac SCs are present in the myocardium in trace amounts: c-kit^{+} – 1 per 10^{4} myocytes [32], Islet-1^{+} – 500-600 in the heart of a 1-5-day-old rats [38], SP^{+} (Sca1) – 500-1000 in the adult heart [39].

Nevertheless, in spite the discovery of the resident CSCs in the heart of mammals, including humans, and accumulation of knowledge about their biology, there is no consensus in the scientific community about the regenerative potential of the myocardium. To date, there is no unifying theory that can explain the fact that the ability of mammalian myocardium to regenerate after injury is lost in a few days after birth [42], and also why body's own resident CSCs, as well as externally injected CSCs [43–46], or mesenchymal SCs [47–49], or bone marrow SCs [50] are unable to form mature cardiomyocytes in the infarct zone. Answers to these questions, on one hand, will establish a real picture of the self-renewal of myocardial cells in normal conditions and changes in homeostasis in injured heart in various cardiovascular diseases, and on the other hand, will make possible to develop a considered tactics to influence on regenerative processes using modulators of proliferation of the resident cardiac cells.

3. Intracellular development of stem cells: a new step in the understanding of biology and behavior of tissue-specific stem cells

From 2005 to 2011, when we studied the behavior of neonatal myocardial cells in primary culture, we confirmed existing concept that after a burst of mitotic activity in the first 2-4 days of postnatal life in vivo [51], the division of neonatal CMs in culture also terminated in 4-5 days after seeding in vitro [52]. At the same time, just like in the body, a mitotic division of 60% of the cells is observed, and after the cessation of this division, a growth of their volume is occurred, that is also similar to the process of hypertrophy of the CMs in vivo. It has been shown that the volume of CMs increased from 819±68 μm^3 on the 1st day to 1532±212 μm^3 on the 3rd day and up to 3246±190 μm^3 on the 6th day of culturing. Moreover, the growth rate of cultured cells is almost completely coincides with the speed of myocyte hypertrophy in the body. Just like in the body, hypertrophy was accompanied by formation of polyploid and multinucleate, mostly binucleate cells. Besides, it was first

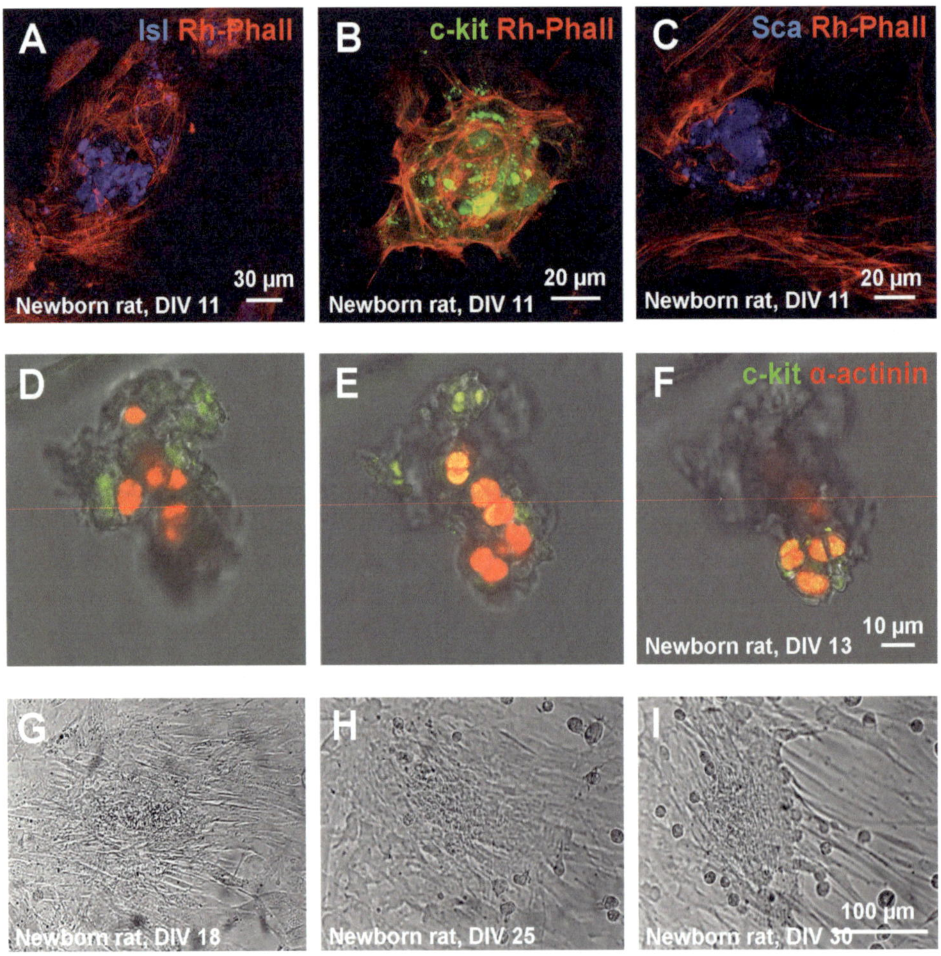

Figure 1. The formation of cardiomyocyte colonies in the primary culture of rat neonatal myocardial cells. (A–C) The optical sections of colonies formed by Isl1⁺, c-kit⁺, and Sca1⁺ CSCs on the 11th day in vitro (DIV). (D-F) Differentiation of c-kit⁺ CSCs inside the colony on the 13th DIV. (G–I) Progressive increase in the contraction rate of the same colony during culturing. (G) The 18th DIV, the contraction rate: 25 beats/min. (H) The 25th DIV, the contraction rate: 46 beats/min. (I) The 30th DIV, the contraction rate: 58 beats/min.

shown that the resident CSCs of three types (c-kit⁺, Sca1⁺, Isl1⁺) in the primary culture of newborn rat myocardial cells form colonies of contracting cardiomyocytes by dividing and subsequent cardiac differentiation, simulating the cardiomyogenesis process from a single immature stem cell to functionally mature cardiomyocyte (**Figure 1**). Therefore, the formation of contractile cardiomyocyte colonies is a complete model for a basic research, testing of drugs and identification of the regenerative potential of CSCs for possible application of the resident self-renewing cells in the treatment of myocardial injury [53, 54]. In addition, immunocytochemical study of the freshly isolated suspension (ex vivo) of myocardial cells from rats of different age groups, adult mice, young bulls and humans, revealed the presence of colonies of different maturity formed by resident CSCs of all three types (**Figure 2**). This made it possible to suggest that the colony formation is a way of myocardial self-renewal that allows to replace the lost cells throughout the life of mammals, including humans.

By study of myocardial cells in monolayer culture from newborn, 20- and 40-day old rats with help of antibodies to CSC antigens and confocal microscopy technique we first revealed the presence of small immature cells (5-6 μm in diameter) of all three types inside mature cardiomyocytes with formation of «cell-in-cell structures» (CICSs). It was found that the cell that has penetrated inside, usually located near the nucleus, or between two nuclei, and stored in the vacuole, membrane of which protects this cell from the host cell cytoplasm (**Figure 3**). Intracellular regions of large cells expressing CSCs antigens increase in size during culturing. Expression of the Ki-67 protein that plays an important role in the regulation of cell division in such vacuoles indicates on proliferation inside located CSCs. As a result of increase in their number, the size of the vacuoles increases, and the membrane thickens and becomes more dense forming a capsule. As time goes by, a formation of two to five openings (pores) occurs in the capsule that appears to provide gas exchange and exchange of metabolites between CSCs and host cell (**Figure 3 C, D, F**). Expression of cardiac (**Figure 3**) proteins (α-sarcomeric actin, sarcomeric α-actinin and Troponin T) suggests that the host cells appear to be a mature cardiomyocytes [55].

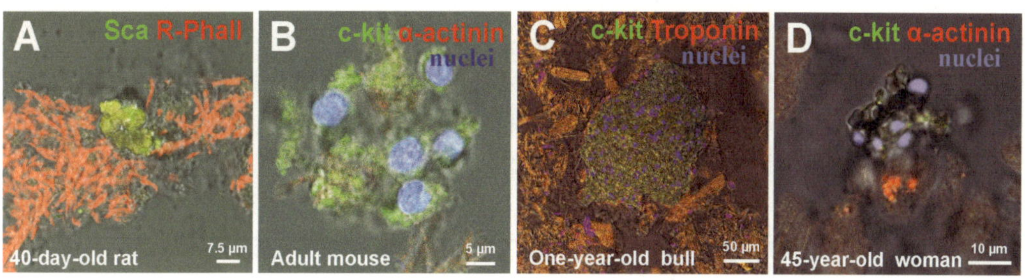

Figure 2. CSC-derived colonies in freshly isolated myocardial cell suspensions (ex vivo) of different mammalian species. The nuclei of the cells have been stained with Hoechst. Transmitted light and fluorescent images are merged. Cardiomyogenic differentiation is verified with expression of sarcomeric α-actinin and troponin T. (A) An undifferentiated small Sca1⁺-colony inside a fragment of myocardium of 40-day-old rat. (B) An undifferentiated small c-kit⁺-colony from myocardium of an adult mouse. (C) A large volume c-kit⁺-colony inside a fragment of myocardium of one-year-old bull. (D) An undifferentiated small c-kit⁺-colony inside a fragment of myocardium of 45-year-old woman.

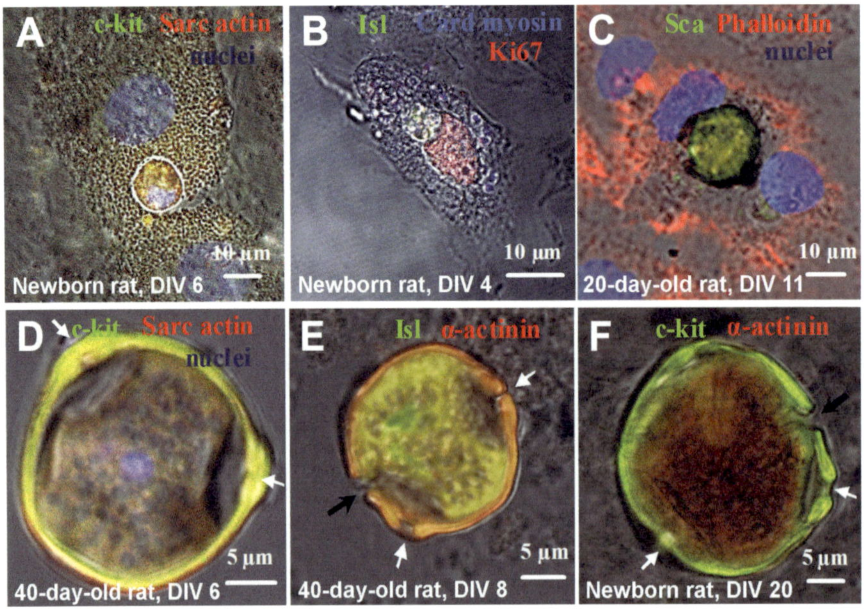

Figure 3. The CSCs inside CMs and the formation of CSC-containing CICSs in the cultures obtained from newborn and 20- and 40-day-old rats. (A–C) Early stages of CICS formation. (A) c-kit$^+$ CSC inside a newborn rat CM obtained (day in vitro 6). (B) Isl1$^+$ CSC inside a newborn rat CM (day in vitro 4). Both the CSC and the host cell exhibit proliferative ability, as documented by the expression of Ki67. (C) Sca + CSC encapsulated between the nuclei of the host cell (20-day-old rat, day in vitro 11). (D–F) The CICS capsule in detail. (D) c-kit$^+$ CSC-containing CICS with a prominent coating ("capsule") with three pores (white arrows) obtained from a newborn rat, day in vitro 6. (E) Erosion (black arrow) of the Isl1$^+$ CSC-containing CICS capsule obtained from 40-day-old rat, day in vitro 8. The pores are also visualized (white arrows). The capsule interior is positive for sarcomeric α-actinin. (F) Erosion (black arrow) of the c-kit$^+$ CSC-containing CICS capsule obtained from a newborn rat, day in vitro 20. The pores are seen (white arrows). The nuclei of the cells have been stained with Hoechst, transmitted light and fluorescent images are merged.

Long-term observation of CICSs in primary culture of newborn, 20- and 40-day old rats revealed that intracellular development of CSCs ended by rupture of CICS-capsule and release of large quantities (up to 200) of transient amplifying cells (TACs), positive against SC-proteins of one of the three CSC-subtypes investigated by us and cardiac proteins (**Figure 4**). In the process of further development of descendants of the original CSC, expression of the stem antigens is reduced, but at the same time, synthesis of cardiac proteins is increased which allowed to suggest a reasonable expectation that the TACs appears to be an intermediate stage between CSCs and cardiomyocytes.

The presence of the similar CICSs, intact and opened, and the TACs of varying degrees of maturity in suspension of myocardial cells (ex vivo) of rats in different ages from the 1st day of life (Figure 4) to old age (1.5 years), adult C57bl/6 mice, 1-year-old bull, and human (45 years old) allowed to suggest that self-renewal of the myocardium may occur by proliferation and initial differentiation of CSCs inside mature cardiomyocytes throughout the life of mammals including humans.

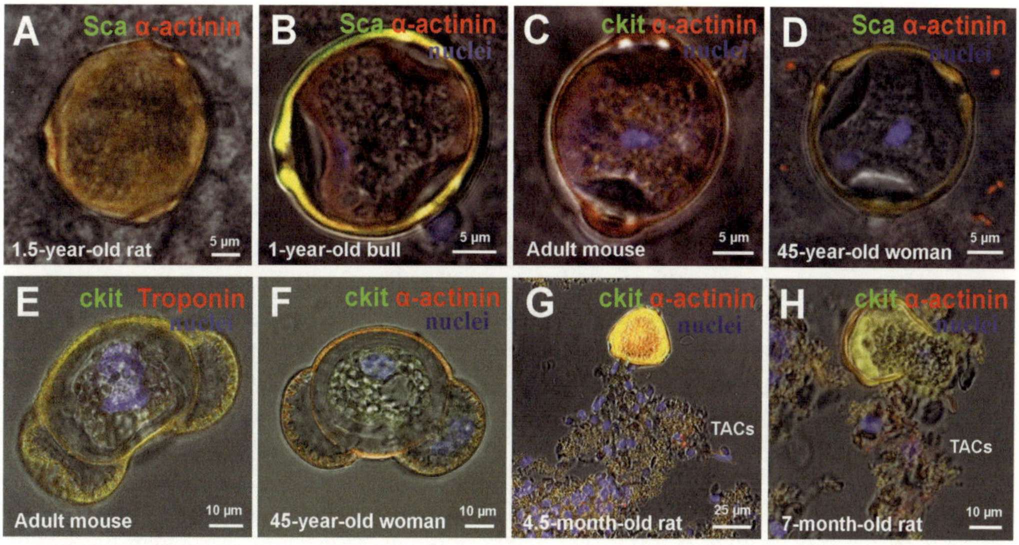

Figure 4. Identification of "cell-in-cell structures" (CICSs) in freshly isolated suspension of myocardial cells (ex vivo) of mammals of different species. (A–D) Intact CICSs. (A) Sca1+ CICS from myocardium of 1.5-year-old rat. (B) $Sca1^+$ CICS from myocardium of 1-year-old bull. (C) c-kit^+ CICS from myocardium of an adult mouse. (D) $Sca1^+$ CICS from myocardium of 45-year-old woman. (E and F) Opened CICSs. (E) c-kit^+ CICS from myocardium of an adult mouse. (F) c-kit^+ CICS from myocardium of 45-year-old woman. (G and H) The release of transitory amplifying cells (TACs) from opened CICSs. (G) Opened CICS and TACs from myocardium of 4.5-month-old rat. (H) Opened CICS and TACs from myocardium of 7-month-old rat.

Because the colonies of contracting cardiomyocytes and intracellular structures are formed in the culture with almost the same frequency (1-2 cases per 100,000 cells plated), we have put forward a hypothesis that the formation of new cardiac cells from resident CSCs may occur by colony formation as well as by intracellular development in mature myocardial cells. It is suggested, therefore, that both processes are present in a healthy heart and, apparently, maintain homeostasis of the heart muscle by replacing the dead cells with new functionally active cardiomyocytes.

It has been found for the first time that in experiments in vitro under simulated conditions of myocardial ischemia zone like acidosis and hypoxia suppress the growth of colonies by blocking the differentiation of resident CSCs, but at the same time, stimulate their intracellular development, increasing the amount of intracellular structures by several times. Therefore, it has been suggested that Ca^{2+}-overload, as well as acidosis and hypoxia, accompanying myocardial infarction, cause the death of not only the mature myocardial cells, but also of the resident CSCs that are forced to «hide» inside large cardiac cells, which allows us to understand their passive role during myocardial infarction. This also explains irrationality of stimulation of resident CSCs for proliferation, differentiation and cell therapy application (introduction of exogenous SCs into the injured region) during the acute phase of ischemia and infarction.

4. Analysis of problems of mammalian myocardial regeneration in the light of the new experimental data

Penetration of one or more cells to another to form the «cell-in-cell structures» (CICSs) was revealed 90 years ago by W.H. Lewis [56] in blood cells. The following terms are used to identify the processes leading to the formation of «cell-in-cell structures»: entosis, emperitosis, cannibalism, emperipolesis and cytophagocytosis. Entosis, firstly described in 2007 [57], is the process of active invasion of one cell (effector cell) into another (host cell). Terms «cannibalism» and «cytophagocytosis» are used to describe the phenomenon of active ingestion by the host cell of another cell which in this case is more passive [58]. Emperipolesis was established in 1956 [59] to indicate the penetration, movement and existence processes of one cell inside the other and, according to Overholtzer and Brugge [58], is suitable to refer to all phenomena in which the formation of structures of «cell-in-cell» type occurs, regardless the fate of cell located inside.

Unlike cytophagocytosis in which dead or dying cells to be destroyed within 30-60 min are captured, if CICS is formed, cell trapped inside nonphagocytic cell can survive for several days or even weeks. But cells in the host cell cytoplasm are surrounded by vacuole membrane that is, presumably, a fragment of a host cell membrane formed after its invagination. Movement of the effector cells, their division, and sometimes exit outside of the host cell were monitored [58].

It is known that the formation of «cell-in-cell structures» occurs both in the lower and in the higher forms of organisms, as well as the penetration of cells of lower forms into cells of highly organized species (parasitism). It is also shown that this phenomenon is typical for immune system cells (cytophagocytosis and emperipolesis) and tumors (entosis), occurring in the inflammation regions [60]. However, there is no information in the scientific literature to date on the formation of such structures in the myocardium.

Discovery of previously unknown mechanism of CSC development within mature cardiac myocytes with subsequent formation of cardiac positive TACs is of great interest not only as another new variant of the «cell-in-cell structure» formation, but also, more significantly, as a phenomenon allowing to reveal the nature of the biological processes that underlie the behavior of CSCs.

Moreover, our in vitro and ex vivo data on the intracellular development of CSCs allow to offer a new perspective into consideration of myocardial infarction problems. The reason is that the current understanding of the physiology of cardiac muscle cells is highly contradictory. For example, data on the rate of formation of new cardiomyocytes range from less than 1% [61–63] to more than 40% per year [64]. In addition, parallel to accumulation of data on the participation of CSCs in cardiomyogenesis, some papers periodically appear arguing in favour of division of mature CMs in the adult heart [64–67], reviving debate about what cells, resident CSCs or mature CMs, participate in the renewal of adult myocardium. In addition, Omatsu-Kanbe et al. [68] found a small population of cells (d~10 μm) that can be differentiated in contracting CMs without preliminary division. In turn, Kimura et al. [69] showed the presence of small dividing cells with characteristics of neonatal CMs: small size, mononuclearity and insignificant oxidative DNA lesions, in the myocardium during hypoxia. Similar neonatal-like cells have been described even after experimental ischemia [70], which allow to suggest that in the myocardium of adult animals these effects stimulate the proliferation of insignificant amount of small cardiac positive cells that are unable, however, to recover the injured myocardium. Therefore, next questions are brought to the fore: what are these cells, where are they located in a healthy heart, and why they are unable to regenerate a myocardium in case of injury? Since the division of large (with a volume of more than 20,000 μm^3) terminally differentiated CMs is contradicted [71], and small dividing cells exceed the size of resident CSCs, a question arises about the origin and the physiological role of these cells.

Proliferative activity of TACs, formed by intracellular proliferation of CSCs, gave us the reason to consider this category of cells as a main regenerative source in the myocardium of adult mammals. We suggest that these cells are able not only to renew the myocardium throughout life, but also partially participate in its regeneration, if not in the acute phase of ischemia and infarction, then possibly during the recovery period or in chronic heart failure.

Thus, discovery of the phenomenon of intracellular development of CSCs is opening further perspectives in the study and solution of the problems of myocardium, and can be seen as a new step in understanding the nature of resident CSCs and development of the approaches for use of cell technologies in regenerative medicine.

Acknowledgements

This work was supported by the Government of Russian Federation (grant no 074-U01), the Russian Foundation of Basic Investigations (projects no 12-04-00941 and no 16-04-01424) and by the state assignment of FASO of Russia ("The mechanisms of development of neuropsychic, metabolic and hormonal dysfunctions in the nervous and endocrine diseases and the approaches for their correction").

Author details

Galina B. Belostotskaya[1,2*], Tatyana A. Golovanova[1,2], Irina V. Nerubatskaya[1,2] and Michael M. Galagudza[2,3]

*Address all correspondence to: gbelost@mail.ru

1 Sechenov Institute of Evolutionary Physiology and Biochemistry of Russian Academy of Sciences, St. Petersburg, Russian Federation

2 Federal Almazov North-West Medical Research Centre, St. Petersburg, Russian Federation

3 ITMO University, St. Petersburg, Russian Federation

References

[1] Maximow A. Der Lymphozyt als gemeinsame Stammzelle der verschiedenen Blutelemente in der embryonalen Entwicklung und im postfetalen Leben der Säugetiere. Originally in: Folia Haematologica. 1909;**8**:125-134. Cell Therapy and Transplantation. 2009,**1**:e.000040.01. DOI: 10.3205/ctt-2008-en-000040.01

[2] Till JE, Mcculloch EA, Siminovitch L. A stochastic model of stem cell proliferation, based on the growth of spleen colony-forming cells. Proceedings of the National Academy of Sciences of the United States of America. 1964;**51**:29-36. PMID: 14104600

[3] Friedenstein AJ, Petrakova KV, Kuralesova AI, Frolova OF. Heterotopic transplants of bone marrow. Analysis of precursor cells for osteogenic and hemopoietic tissues. Transplantation. 1968;**6**:230-247. PMID: 5654088

[4] Potten CS, Hendry JH. Clonogenic cells and stem cells in epidermis. International Journal of Radiation Biology and Related Studies in Physics, Chemistry, and Medicine. 1973;**24**:537-540. PMID: 4543437

[5] Tseng SCG. Concept and application of limbal stem cells. Eye. 1989;**3**:141-157. DOI: 10.1038/eye.1989.22

[6] Dekel B, Zangi L, Shezen E, Reich-Zeliger S, Eventov-Friedman S, Katchman H, Jacob-Hirsch J, Amariglio N, Rechavi G, Margalit R, Reisner Y. Isolation and characterization of nontubular Sca-1+Lin- multipotent stem/progenitor cells from adult mouse kidney. Journal of the American Society of Nephrology. 2006;**17**:3300-3314. DOI: 10.1681/ASN.2005020195

[7] Knight B, Matthews VB, Olynyk JK, Yeoh GC. Jekyll and Hyde: Evolving perspectives on the function and potential of the adult liver progenitor (oval) cell. Bioessays. 2005;**27**:1192-1202. DOI: 10.1002/bies.20311

[8] Gronthos S, Mankani M, Brahim J, Robey PG, Shi S. Postnatal human dental pulp stem cells (DPSCs) in vitro and in vivo. PNAS. 2000;**97**:13625-13630. DOI: 10.1073/pnas.240309797

[9] Mao X, Liu Y, Chen C, Shi S. Mesenchymal stem cells and their role in dental medicine. Dental Clinics of North America. 2017;**61**:161-172. DOI: 10.1016/j.cden.2016.08.006

[10] Altman J. Are new neurons formed in the brains of adult mammals? Science. 1962;**135**:1127-1128. PMID: 13860748

[11] Nottebohm F. Neuronal replacement in adulthood. Annals of the New York Academy of Sciences. 1985;**457**:143-161. PMID: 3913361

[12] Chetverukhin VK, Polenov AL. Ultrastructural radioautographic analysis of neurogenesis in the hypothalamus of the adult frog, *Rana temporaria*, with special reference to physiological regeneration of the preoptic nucleus. I. Ventricular zone cell proliferation. Cell and Tissue Research. 1993;**271**:341-350. PMID: 8453658

[13] Polenov AL, Chetverukhin VK. ltrastructural radioautographic analysis of neurogenesis in the hypothalamus of the adult frog, *Rana temporaria*, with special reference to physiological regeneration of the preoptic nucleus. II. Types of neuronal cells produced. **Cell** and Tissue Research. 1993;**271**:351-362. PMID: 8453659

[14] Frederiksen K, McKay RD. Proliferation and differentiation of rat neuroepithelial precursor cells in vivo. Journal of Neuroscience. 1988;**8**:1144-1151. PMID: 3357014

[15] Alvarez-Buylla A, Lois C. Neuronal stem cells in the brain of adult vertebrates. Stem Cells. 1995;**13**:263-272. DOI: 10.1002/stem.5530130307

[16] Sullivan R, Dailey T, Duncan K, Abel N, Borlongan CV. Peripheral nerve injury: Stem cell therapy and peripheral nerve transfer. International Journal of Molecular Sciences. 2016;**17**:2101-2112. DOI: 10.3390/ijms17122101

[17] Oh SK, Jeon SR. Current concept of stem cell therapy for spinal cord injury: A review. Korean Journal of Neurotrauma. 2016;**12**:40-46. DOI: 10.13004/kjnt.2016.12.2.40

[18] Duan H, Song W, Zhao W, Gao Y, Yang Z, Li X. Endogenous neurogenesis in adult mammals after spinal cord injury. Science China Life Sciences. 2016;**59**:1313-1318. DOI: 10.1007/s11427-016-0205-2

[19] Jessberger S. Stem cell-mediated regeneration of the adult brain. Transfusion Medicine and Hemotherapy. 2016;**43**:321-326. DOI: 10.1159/000447646

[20] Franklin RJ, French-Constant C. Remyelination in the CNS: From biology to therapy. Nature Reviews Neuroscience. 2008;**9**:839-855. DOI: 10.1038/nrn2480.

[21] Lee JH, Oh IH, Lim HK. Stem cell therapy: A prospective treatment for Alzheimer's disease. Psychiatry Investigation. 2016;**13**:583-589. DOI: 10.4306/pi.2016.13.6.583

[22] Kremer D, Gottle P, Hartung HP, Kury P. Pushing forward: Remyelination as the new frontier in CNS diseases. Trends in Neurosciences. 2016;**39**:246-263. DOI: 10.1016/j.tins.2016.02.004

[23] Leri A, Kajstura J, Anversa P. Cardiac stem cells and mechanisms of myocardial regeneration. Physiological Reviews. 2005;**85**:1373-1416. DOI: 10.1152/physrev.00013.2005

[24] Rumyantsev PP. Interrelations of the proliferation and differentiation processes during cardiac myogenesis and regeneration. International Review of Cytology. 1977;**51**:187-273. PMID: 338537

[25] Rumyantsev PP, Borisov A. DNA synthesis in myocytes from different myocardial compartments of young rats in norm, after experimental infarction and in vitro. Biomedica Biochimica Acta. 1987;**46**:S610-S615. PMID: 3435518

[26] Erokhina IL, Rumyantsev PP. Proliferation and biosynthetic activities of myocytes from conductive system and working myocardium of the developing mouse heart. Light microscopic autoradiographic study. Acta Histochemica. 1988;**84**:51-66. DOI: 10.1016/S0065-1281(88)80010-2

[27] Kajstura J, Leri A, Finato N, et al. Myocyte proliferation in end-stage cardiac failure in humans. Proceedings of the National Academy of Sciences of the United States of America. 1998;**95**:8801-8805. PMID: 9671759

[28] Beltrami AP, Urbanek K, Kajstura J, et al. Evidence that human cardiac myocytes divide after myocardial infarction. New England Journal of Medicine. 2001;**344**:1750-1757. DOI: 10.1056/NEJM200106073442303

[29] Leri A, Barlucchi L, Limana F, et al. Telomerase expression and activity are coupled with myocyte proliferation and preservation of telomeric length in the failing heart. Proceedings of the National Academy of Sciences of the United States of America. 2001;**98**:8626-8631. DOI: 10.1073/pnas.151013298

[30] Quaini F, Urbanek K. Beltrami AP, et al. Chimerism of the transplanted heart. New England Journal of Medicine. 2002;**346**:5-15. DOI: 10.1056/NEJMoa012081

[31] Hierlihy AM, Seale P, Lobe CG, et al. The post-natal heart contains a myocardial stem cell population. FEBS Letters. 2002;**530**:239-243. PMID: 12387899

[32] Beltrami AP, Barlucchi L, Torella D, et al. Adult cardiac stem cells are multipotent and support myocardial regeneration. Cell. 2003;**114**:763-776. PMID: 14505575

[33] Urbanek K, Quaini F, Tasca G, et al. Intense myocyte formation from cardiac stem cells in human cardiac hypertrophy. Proceedings of the National Academy of Sciences of the United States of America. 2003;**100**:10440-10445. DOI: 10.1073/pnas.1832855100

[34] Cai C-L, Liang X, Shi Y, et al. Isl1 identifies a cardiac progenitor population that proliferates prior to differentiation and contributes a majority of cells to the heart. Developmental Cell. 2003;**5**:877-889

[35] Anversa P, Kajstura J, Leri A, Bolli R. Life and death of cardiac stem cells: A paradigm shift in cardiac biology. Circulation. 2006;**113**:1451-1463. DOI: 10.1161/CIRCULATIONAHA.105.595181

[36] Oh H, Bradfute SB, Gallardo TD, et al. Cardiac progenitor cells from adult myocardium: Homing, differentiation, and fusion after infarction. PNAS. 2003;**100**:12313-12318. DOI: 10.1073/pnas.2132126100

[37] Matsuura K, Nagai T, Nishigaki N, et al. Adult cardiac Sca-1-positive cells differentiate into beating cardiomycytes. Journal of Biological Chemistry. 2004;**279**:11384-11391. DOI: 10.1074/jbc.M310822200

[38] Laugwitz K-L, Moretti A, Lam J, et al. Postnatal isl1+ cardioblasts enter fully differentiated cardiomyocyte lineages. Nature. 2005;**433**:647-653. DOI: 10.1038/nature03215

[39] Pfister O, Mouquet F, Jain M, et al. CD31− but not CD31+ cardiac side population cells exhibit functional cardiomyogenic differentiation. Circulation Research. 2005;**97**:52-61. DOI: 10.1161/01.RES.0000173297.53793.fa

[40] van Vliet P, Roccio M, Smits AM, et al. Progenitor cells isolated from the human heart: A potential cell source for regenerative therapy. Netherlands Heart Journal. 2008;**16**:163-169. PMID: 18566670

[41] Belostotskaya GB, Golovanova TA. Characterization of contracting cardiomyocyte colonies in the primary culture of neonatal rat myocardial cells: A model of in vitro cardiomyogenesis. Cell Cycle. 2014;**13**:910-918. DOI: 10.4161/cc.27768

[42] Porrello ER, Mahmoud AI, Simpson E, et al. Transient regenerative potential of the neonatal mouse heart. Science. 2011;**331**:1078-1080. DOI: 10.1126/science.1200708

[43] Centola M, Schuleri KH, Lardo AC, Hare JM. Stem cell therapy for myocardial regeneration: Mechanisms and current clinical applications. Giornale Italiano Di Cardiologia (Rome). 2008;**9**:234-250. PMID: 18543793

[44] Smits AM, van Laake LW, den Ouden K, et al. Human cardiomyocyte progenitor cell transplantation preserves long-term function of the infarcted mouse myocardium. Cardiovascular Research. 2009;**83**:527-535. DOI: 10.1093/cvr/cvp146

[45] D'Alessandro DA, Kajstura J, Hosoda T, et al. Progenitor cells from the explanted heart generate immunocompatible myocardium within the transplanted donor heart. Circulation Research. 2009;**105**:1128-1140. DOI: 10.1161/CIRCRESAHA.109.207266

[46] Welt FG, Gallegos R, Connell J, et al. Effect of cardiac stem cells on left ventricular remodeling in a canine model of chronic myocardial infarction. Circulation: Heart Failure. 2013;**6**:99-106. DOI: 10.1161/CIRCHEARTFAILURE.112.972273

[47] Seth S, Narang R, Bhargava B, et al. Percutaneous intracoronary cellular cardiomyoplasty for nonischemic cardiomyopathy: Clinical and histopathological results: The first-in-man ABCD (Autologous Bone Marrow Cells in Dilated Cardiomyopathy) trial. Journal of the American College of Cardiology. 2006;**48**:2350-2351. DOI: 10.1016/j.jacc.2006.07.057

[48] Vrtovec B, Poglajen G, Lezaic L, et al. Effects of intracoronary CD34+ stem cell transplantation in nonischemic dilated cardiomyopathy patients: 5-year follow-up. Circulation Research. 2013;**112**:165-173. DOI: 10.1161/CIRCRESAHA.112.276519

[49] Fischer-Rasokat U, Assmus B, Seeger FH, et al. A pilot trial to assess potential effects of selective intracoronary bone marrow-derived progenitor cell infusion in patients with nonischemic dilated cardiomyopathy: Final 1-year results of the transplantation of progenitor cells and functional regeneration enhancement pilot trial in patients with nonischemic dilated cardiomyopathy. Circulation: Heart Failure. 2009;**2**:417-423. DOI: 10.1161/CIRCHEARTFAILURE.109.855023

[50] Wei HM, Wong P, Hsu LF, Shim W. Human bone marrow-derived adult stem cells for post-myocardial infarction cardiac repair: Current status and future directions. Singapore Medical Journal. 2009;**50**:935-942

[51] Golovanova TA, Belostotskaya GB. Regeneration of rat cardiac myocytes in vitro: The proliferative activity of neonatal cardiac myocytes. Cell Transplantology and Tissue Engineering. 2011;**6**:66-70 (in Russian)

[52] Li E, Wang X, Capasso J, Gerdes A. Rapid transition of cardiac myocytes from hyperplasia to hypertrophy during postnatal development. Journal of Molecular and Cellular Cardiology. 1996;**28**:1737-1746. DOI: 10.1006/jmcc.1996.0163

[53] Golovanova TA, Belostotskaya GB. Regeneration of rat cardiac myocytes in vitro: Contracting myocyte colonies in the primary culture of neonatal cardiac myocytes. Cell Transplantology and Tissue Engineering. 2012;**7**:67-72. (in Russian)

[54] Belostotskaya GB, Golovanova TA. Characterization of contracting cardiomyocyte colonies in the primary culture of neonatal rat myocardial cells: A model of in vitro cardiomyogenesis. Cell Cycle. 2014;**13**:910-918. DOI: 10.4161/cc.27768

[55] Belostotskaya G, Nevorotin A, Galagudza M. Identification of cardiac stem cells within mature cardiac myocytes. Cell Cycle. 2015;**14**:3155-3162. DOI: 10.1080/15384101.2015.1078037

[56] Lewis WH. The exgulfment of living blood cells by others of the same type. The Anatomical Record. 1925;**3**:43-49

[57] Overholtzer M, Mailleux AA, Mouneimne G, et al. A non-apoptotic cell death process, entosis, that occurs by cell-in-cell invasion. Cell. 2007;**131**:966-979. DOI: 10.1016/j.cell.2007.10.040

[58] Overholtzer M, Brugge JS. The cell biology of cell-in-cell structures. Nature Reviews Molecular Cell Biology. 2008;**9**:796-809. DOI: 10.1038/nrm2504

[59] Humble JG, Jayne WH, Pulvertaft R.J. Biological interaction between lymphocytes and other cells. British Journal of Haematology. 1956;**2**:283-294. DOI: 10.1111/j.1365-2141.1956.tb06700.x

[60] He MF, Wang S, Wang Y, Wang XN. Modeling cell-in-cell structure into its biological significance. Cell Death and Disease. 2013;**4**:e630. DOI: 10.1038/cddis.2013.147

[61] Soonpaa MH, Koh GY, Pajak L, et al. Cyclin D1 overexpression promotes cardiomyocyte DNA synthesis and multinucleation in transgenic mice. Journal of Clinical Investigation. 1997;**99**:2644-2654. DOI: 10.1172/JCI119453

[62] Bergmann O, Bhardwaj RD, Bernard S, et al. Evidence for cardiomyocyte renewal in humans. Science. 2009;**324**:98-102. DOI: 10.1126/science.1164680

[63] Walsh S, Ponten A, Fleischmann B, Jovinge S. Cardiomyocyte cell cycle control and growth estimation in vivo—An analysis based on cardiomyocyte nuclei. Cardiovascular Research. 2010;**86**:365-373. DOI: 10.1093/cvr/cvq005

[64] Kajstura J, Gurusamy N, Ogórek B, et al. Myocyte turnover in the aging human heart. Cardiovascular Research 2010;**107**:1374-1386. DOI: 10.1161/CIRCRESAHA.110.231498

[65] Leri A, Kajstura J, Anversa P. Mechanisms of myocardial regeneration. Trends in Cardiovascular Medicine. 2011;**21**:52-58. DOI: 10.1016/j.tcm.2012.02.006

[66] Senyo SE, Steinhauser ML, Pizzimenti CL, et al. Mammalian heart renewal by pre-existing cardiomyocytes. Nature. 2013;**493**:433-436. DOI: 10.1038/nature11682

[67] Mollova M, Bersell K, Walsh S, et al. Cardiomyocyte proliferation contributes to heart growth in young humans. Proceedings of the National Academy of Sciences of the United States of America. 2013;**110**:1446-1451. DOI: 10.1073/pnas.1214608110

[68] Omatsu-Kanbe M, Matsuura H. A novel type of self-beating cardiomyocytes in adult mouse ventricles. Biochemical and Biophysical Research Communications. 2009;**381**:361-366. DOI: 10.1016/j.bbrc.2009.02.048

[69] Kimura W, Xiao F, Canseco DC, et al. Hypoxia fate mapping identifies cycling cardiomyocytes in the adult heart. Nature. 2015;**523**:226-230. DOI: 10.1038/nature14582

[70] Tang X-L, Li Q, Rokosh G, et al. Long-Term Outcome of administration of c-kitPOS cardiac progenitor cells after acute myocardial infarction: Transplanted cells do not become cardiomyocytes, but structural and functional improvement and proliferation of endogenous cells persist for at least one year. Circulation Research. 2016;**118**:1091-1105. DOI: 10.1161/CIRCRESAHA.115.307647

[71] Leri A, Rota M, Pasqualini FS, et al. Origin of cardiomyocytes in the adult heart. Circulation Research. 2015;**116**:150-166. DOI: 10.1161/CIRCRESAHA.116.303595

Chapter 6

The New Pharmacological Approaches for the Regulation of Functional Activity of G Protein-Coupled Receptors

Alexander O. Shpakov and Kira V. Derkach

Additional information is available at the end of the chapter

http://dx.doi.org/10.5772/intechopen.73322

Abstract

The G protein-coupled receptors (GPCRs), a large family of the receptors that specifically interact with a number of signal molecules, play a key role in the regulation of fundamental cell processes, and the pharmacological action of over 40% of drugs is carried out through GPCRs. In the last years, a significant progress was made in the creation of selective regulators of GPCRs interacting with their allosteric sites, such as the synthetic peptides corresponding to intracellular regions of receptors (GPCR-peptides) and the low-molecular weight agonists and antagonists of GPCRs. This review describes the recent results obtained by us and other authors in the development of GPCR-peptides and low-molecular weight agonists and the prospects of their use in clinics.

Keywords: low-molecular weight agonist, G protein-coupled receptor, allosteric regulation, GTP-binding protein, thienopyrimidine

1. Introduction

The G protein-coupled receptors (GPCRs) are a large family of cell surface receptors that specifically interact with a number of signal molecules, such as amino acids, nucleotides, peptides, proteins, lipids and odorants. GPCRs play a key role in the regulation of fundamental cell processes including growth, metabolism, differentiation, motility and apoptosis [1, 2]. The evolution of GPCRs has about 700 million years, and they are characterized by the similarity of the structural and functional organization and membrane topology [3]. GPCRs contain seven transmembrane regions (TM) forming a transmembrane channel, N-terminal extracellular and C-terminal intracellular domains (CTD) and three extracellular and three intracellular loops (ICLs). The ICLs of agonist-bound GPCR interact with $\alpha\beta\gamma$-heterotrimeric GTP-binding proteins (G proteins) and β-arrestins, which regulate activity of adenylyl cyclase (AC), phospholipase C, phosphatidylinositol 3-kinase, mitogen-activated protein kinases and G protein-gated ion channels. In most cases, the membrane-proximal regions of the second and third ICLs are involved in the interaction with G proteins, while the third ICL and CTD with β-arrestins. The extracellular regions are responsible for ligand recognition and participate in the formation of high-affinity ligand-binding (orthosteric) site, which is usually located in the transmembrane channel of GPCRs and, in certain receptors, is placed in a large ectodomain.

The action of over 40% of the currently used drugs affecting physiological and biochemical processes is carried out through GPCRs [4]. The changes in GPCRs and their signaling cascades lead to a large number of diseases, indicating that the search of new selective and effective regulators of GPCRs is the critical challenge to medical endocrinology and biochemistry. Nowadays, one of the most widely used approaches to create GPCR regulators is the development of synthetic analogs of natural hormones, which specifically interact with the orthosteric site of GPCR and having high efficiency and selectivity. This approach began to give very good results when the 3D structure for a number of GPCRs was established using X-ray analysis of their crystal forms [5]. In the last years, a significant progress was made in the development of GPCR regulators interacting with the allosteric sites of receptor. For this purpose, a few strategies can be used, and among them, greatest interests are the design of peptides corresponding to intracellular regions of GPCRs (GPCR-peptides) and the creation of low-molecular weight (LMW) agonists and antagonists. The GPCR-peptides and LMW regulators specifically bind to the allosteric sites of receptor located in the ICLs and in the transmembrane domain, respectively.

This review describes the recent results obtained by us and other authors in the development of GPCR-peptides and LMW agonists and the prospects of their use in clinic.

2. The GPCR-peptides and their lipophilic derivatives

Currently, a lot of data were obtained that synthetic peptides corresponding to intracellular regions of GPCRs specifically influence the activity of cognate receptors and intracellular pathways dependent on them [6–9]. Since the regions interacting with G proteins and β-arrestins are located primarily in membrane-proximal segments of ICLs, conformation of these regions in full-size GPCR is stabilized by hydrophobic segments of TM and by interactions between N- and C-termini of ICLs. Therefore, the modification of GPCR-peptides by hydrophobic radicals and lipophilic amino acid sequences simulating TM, as well as the cyclization of GPCR-peptides should lead to significant increase in biological activity of modified GPCR-peptides, which was confirmed by the *in vivo* and the *in vitro* experiments [6, 8, 10, 11]. GPCR-peptides modified by hydrophobic radicals and amino acid sequences, which are similar in physicochemical properties to TM segments, are able to penetrate the plasma membrane, to incorporate into the TM/ICL interfaces formed by membrane-proximal segments of

ICLs and cytoplasmic portion of TM, and to interact with the allosteric sites of cognate GPCR (**Figure 1**). The GPCR-peptides modified by hydrophobic radicals, primarily by palmitoyl and myristoyl radicals, are designated as pepducins [6]. They influence the signal transduction, acting as intracellular allosteric agonists or antagonists, and are capable of triggering the appropriate cell response in the absence of hormonal stimulus [12–16]. As pepducins specifically interact with complementary regions of cognate GPCR, their effects are receptor specific. Pepducins do not affect even the closely related receptors and are active only in tissues and cells where the receptors homologous to them are expressed [17, 18].

Lipophilic derivatives of GPCR-peptides corresponding to ICLs of types 1, 2 and 4 protease-activated receptors (PARs) influence platelet aggregation, inflammation, angiogenesis, apoptosis, transformation and metastasis [19]. Pepducin P1pal-7 (PZ-128), a derivative of N-terminal segment of the third ICL (ICL3) of PAR1, functions as antagonist and significantly inhibits PAR1-mediated platelet aggregation and arterial thrombosis, being a good alternative to LMW PAR1-antagonists used to treat arterial thrombosis. The compound PZ-128 suppresses the tumor growth and metastasis, reduces the cell viability in breast, ovarian and lung carcinoma cells, inhibits PAR1-mediated migration of lung cancer cells and decreases lung tumor growth [20, 21]. The efficiency of PZ-128 is comparable to that of Bevacizumab, a widely used antitumor agent and angiogenesis inhibitor. Lipophilic derivatives of GPCR-peptides corresponding to ICLs of the chemokine receptors CXCR1, CXCR2 and CXCR4 also possess the potent antitumor activity [12, 13]. The CXCR4-derived pepducins with high efficiency

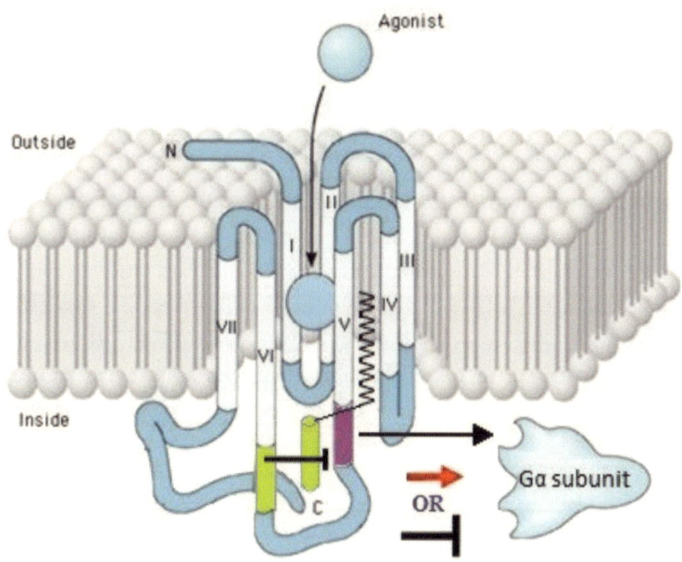

Figure 1. The molecular mechanisms of action of GPCR-peptides modified by hydrophobic radicals. Specific interaction of GPCR-peptide (in green) with complementary region (in purple) of the cognate receptor leads to: (1) the disconnection of the interaction between receptor region (in green) homologous to GPCR-peptide with complementary region; (2) the changes in conformation of complementary region that is involved in the interaction with G protein α-subunit; and, finally, (3) the activation or, on the contrary, inhibition of G protein α-subunit and downstream effector enzymes.

suppress the survival and metastasis of disseminated lymphoma xenografts, which can be the basis of their application in treating lymphoid malignancies [13]. The pepducin P2pal-18S corresponding to the ICL3 of PAR2, an allosteric antagonist of this receptor, and its analogs protect acinar cells against injury induced by bile acid and reduce the severity of acute biliary pancreatitis in mice, which indicates a good opportunity for their use in management of patients with high risk of acute biliary pancreatitis [22].

We synthesized and studied a large number of lipophilic GPCR-peptides that correspond to the ICL3 of type 1 relaxin receptor (RXFP1), 5-hydroxytryptamine receptor of the type 6 (5-HT_6R) and the receptors of luteinizing hormone/chorionic gonadotropin (LH/ChG-R) and thyroid-stimulating hormone (TSHR). These receptors are involved in the regulation of nervous, endocrine and reproductive systems. Binding of these receptors to hormones induces activation of G_s proteins and the enzyme adenylyl cyclase (AC), which leads to increase in intracellular cyclic adenosine monophosphate (cAMP) level and activation of cAMP-dependent enzymes and transcriptional factors. The choice of the ICL3 for synthesis of GPCR-peptides was due to the fact that this loop is responsible for specific binding and activation of G proteins [23].

The C-palmitoylated 11-mer peptide QVKKE(Nle)ILAKR$^{619–629}$K(Pal) corresponding to C-terminal portion of the ICL3 of RXFP1 stimulated AC activity and GTP-binding capacity of G_s proteins and, in addition, reduced regulatory effects of relaxin. The influence of 11-mer palmitoylated peptide on the AC signaling system (ACSS) was more pronounced as compared with 15-mer peptide 615–629 lacking fatty acid radical, while unmodified 11-mer peptide was not active [14]. The C-palmitoylated peptide KHSRKALKASL$^{258–268}$K(Pal)A corresponding to the ICL3 of 5-HT_6R stimulated the activity of G_s proteins and AC much more effectively than its unmodified analog [15]. The similar results were obtained for PAR1-derived peptides. Peptide PAR1-295-313, lacking hydrophobic radical, had a little effect on activity of calcium channels, while its palmitoylated analog caused a rapid stimulation of Ca^{2+} influxes [6]. The regulatory effects of peptides 619-629-K(Pal) and 258-268-K(Pal)A were tissue specific. The peptides 619-629-K(Pal) stimulated AC in plasma membranes isolated from the myocardium and brain that are enriched by the receptors RXFP1, but had no effect in the skeletal muscles where there are no RXFP1 [14]. Palmitoylated 5-HT_6R-peptide was effective in the brain rich in 5-HT_6R, but not in the myocardium and testes where 5-HT_6R are absent or expressed very little [15]. These results give grounds to make a conclusion that modification of GPCR-peptides by hydrophobic radicals is one of the most perspective approaches to enhance their specific biological activity.

In the *in vitro* experiments, the lysine-palmitoylated peptide 612-627-K(Pal)A, the derivatives of C-terminal portion of the ICL3 of TSHR, stimulated the basal AC activity and GTP binding of G_s proteins in the thyroidal membranes and reduced the AC activity stimulated by TSH [16]. In the *in vivo* experiments, single and 3-day treatment of rats with intranasally or intramuscularly administered peptide 612-627-K(Pal)A led to increase of thyroid hormones level and, in the case of 3-day treatment, to decrease of plasma TSH concentration. Two hours after a single intranasal administration of peptide 612-627-K(Pal)A (450 μg/kg), the level of free thyroxine (fT_4) and total 3,5,3′-triiodothyronine (tT_3) was increased by 31 and 37%. The stimulating effect of 612-627-K(Pal)A on thyroid function was enhanced on the second day of treatment and then weakened, which was due to the decrease of thyroid sensitivity to the peptidic TSHR agonist. The evidence for this was the weakening of stimulating effect of intranasally administered thyrotropin-releasing hormone on the production of thyroid hormones in rats treated for 2 days with 612-627-K(Pal)A [11]. Unmodified peptide 612-627-KA possessed a low activity in the *in vitro* experiments and had no influence on thyroid status of animals irrespective of the mode of administration [11, 16]. Thus, the ability of peptide 612-627-K(Pal)A to stimulate the production of thyroid hormones gives grounds to consider it as a prototype for creating the novel thyroid regulators.

To estimate the effect of localization and length of fatty acid radicals on biological activity of GPCR-peptides, we synthesized the series of acylated analogs of peptide 562–572 corresponding to the ICL3 of LH/ChG-R and studied their influence on gonadotropin-sensitive ACSS in rat testicular membranes. The lipophilic derivatives of peptide 562–572 containing palmitoyl and decanoyl radicals at the N- or C-terminus, or at both termini, were investigated. We showed that lipophilic peptides modified by fatty acid radicals at the C-terminus, where in full-size LH/ChG-R the sixth TM is located, stimulated in a dose-dependent manner the basal AC activity and GppNHp binding of G_s proteins and reduced the AC stimulating effect of human ChG [24]. The C-palmitoylated peptide 562–572 was much more active than its decanoyl counterpart. The N-acylated peptides had no effect on the ACSS. These data support the view that hydrophobic radical in GPCR-peptides mimics TM and must be comparable with it in the size and hydrophobicity. The establishment of the criteria for modification of GPCR-peptides with hydrophobic radicals is one of the most promising ways to create drugs capable of regulating the biochemical and physiological processes *in vivo*.

The progress achieved in development of GPCR-peptides open up prospects for their wide application in pharmacology and medicine as drugs to treat cancer, immunological, endocrine, cardiovascular and other diseases, as well as in fundamental biology as an instrument to study the molecular mechanisms of GPCR interaction with ligands and intracellular regulatory and effector proteins. The modification of GPCR-peptides can significantly change their selectivity, efficiency, bioavailability and stability, which allows unlimited expand the field of the use of the peptides. Note, in the last years, the works were carried out on the development of peptides corresponding to functionally active regions of receptor tyrosine kinases, G proteins and the enzymes generating the second messengers [7, 9].

3. The low-molecular weight (LMW) allosteric regulators of receptors of glycoprotein hormones

Unlike most of the other GPCRs, LH/ChG-R and TSHR have a large N-terminal extracellular domain (ectodomain), which forms orthosteric site for high-affinity binding of glycoprotein hormones [25, 26]. Typically, high-affinity binding of ligand occurs in the orthosteric site located in transmembrane channel of receptor. The transmembrane channel of LH/ChG-R and TSHR, on the contrary, involves the allosteric site that remains free when receptor is occupied by hormone. The binding of glycoprotein hormones causes conformational rearrangements in ectodomain and leads to the changes in the interaction between ectodomain and extracellular loops of GPCR. This induces the conformational changes in the transmembrane channel and allosteric site located therein, in ICLs forming G protein-binding surface of receptor and causes activation of G proteins and intracellular effectors, including AC and phosphoinositide-specific phospholipase C (PLC) [26, 27].

The first LMW ligands of LH/ChG-R, the derivatives of thienopyrimidines, were discovered in 2002 [28]. They possess the ability to penetrate into transmembrane channel, specifically interact with the allosteric site located therein and, as a result, activate LH-dependent signaling cascades. Based on the structure of thienopyrimidine derivatives, in the recent years, the LMW agonists and antagonists of TSHR were developed, which opened a new way in the pharmacological treatment of thyroid diseases [29, 30].

The LH and human ChG, the endogenous ligands of LH/ChG-R, are the αβ-heterodimers consisting of highly conservative α-subunit and variable β-subunit, and the β-subunit determines specificity of gonadotropins binding to LH/ChG-R ectodomain. Meanwhile, the use of LH and human ChG in medicine is limited by the rapid development of resistance of reproductive tissues to these hormones, their immunogenicity, and the requirement of parenteral administration of gonadotropins. All this makes it necessary to develop the new LH/ChG-R regulators, which differ in the mechanisms of action from gonadotropins. Searching for such regulators led to the discovery of several classes of chemical compounds, including the pyrazole and 1,3,5-terphenyl derivatives. The most active among them were the thienopyrimidine derivatives, such as compound Org 41,841, *N-tert*-butyl-5-amino-4-(3-methoxyphenyl)-2-(methylthio)thieno[2,3-*d*]pyrimidine-6-carboxamide and its analog Org 43553 [28, 31].

The Org 43553 binds specifically to LH/ChG-R (K_d, 2.4 nM), and the treatment of cells with [^{125}I]-ChG does not cause the dissociation of Org 43553. At nanomolar concentrations, Org 43553 activates AC and cAMP-dependent transcriptional factors with the efficiency of 62 and 80% of that LH, but had a little effect on the activity of receptor of follicle-stimulating hormone and TSHR [31]. It is known that gonadotropins via G_q protein stimulate PLC, and activation of this enzyme requires higher concentrations of LH and human ChG [32]. The treatment of cells with the compound Org 43553 at concentrations of 10^{-6} and 10^{-5} M resulted in an increase of PLC activity to 33–37%, which is less than 5% of the corresponding effect of LH. These data indicate that the binding of LH/ChG-R with Org 43553 has a little effect on ability of this receptor to interact with G_q proteins and activate phosphoinositide turnover [31].

The Org 43553 was active in the *in vivo* conditions, including the oral administration [33, 34]. A single oral administration of Org 43553 at a dose of 50 mg/kg induced the ovulation in immature mice and adult rats. Ovulated oocytes were characterized by high fertility, and their implantation resulted in the formation of normal embryos. Oral administration of Org 43553 at the same dose to adult male rats increased the plasma level of testosterone. The study of pharmacokinetics of Org 43553 showed that, in comparison with gonadotropins, this compound degrades more quickly [33]. A decrease in the half-life has a great practical importance, since it allows reducing the risk of ovarian hyperstimulation syndrome, one of the most severe complications of gonadotropin-stimulated ovulation. A single oral administration of

Org 43553 to female rats had no effect on the size of ovaries and on the permeability of ovarian blood vessels, and even multiple treatments of animals with Org 43553 did not result in the development of ovarian hyperstimulation syndrome [34]. The cause of this is that the compound Org 43553 reduces the level of vascular endothelial growth factor, a crucial inducer of vascular permeability, while gonadotropins increase the concentrations of this factor. At a dose of 300 mg, Org 43553 induced ovulation in 83% of women of reproductive age. In the case of experimental animals, the signs of ovarian hyperstimulation syndrome were not observed, and this fact indicates good prospects for the use of Org 43553 in the induction of ovulation in clinics [35].

We studied new analogs of Org 43553, the compounds TP01 and TP02, which were synthesized by acylation of thienopyrimidine precursor at the 5-amino group using isonicotinoyl and thiophene-3-carbonyl chlorides, respectively. Both compounds were active in the *in vitro* conditions, stimulating ACSS in the reproductive tissues [36, 37], and in the *in vivo*, increasing testosterone level in the case of different mode of drug administration to male rats [38]. The TP01 and TP02 stimulated the basal AC activity in plasma membranes isolated from the testes and ovaries of rats, and the EC_{50} values for their effects on AC activity amounted to 1.05–1.12 mM and 280–365 nM, respectively. Along with this, in testicular membranes, the TP01 and TP02 increased GTP-binding capacity of G_s proteins [36]. The AC stimulating effect of human ChG in the presence of thienopyrimidine derivatives was retained, and at low, nonsaturating concentrations, the additive effect of human ChG and TP01/TP02 was observed, indicating different localization of gonadotropin- and thienopyrimidine-binding sites in LH/ChG-R [37].

In the *in vivo* experiments on rats, the compound TP01 administered intraperitoneally at a dose 15 mg/kg after 3 and 5 h increased the plasma testosterone level by 83 and 339% and at a dose 27 mg/kg—by 134 and 325%, respectively. The increase of testosterone level 3 h after TP01 treatment at the doses 15 and 27 mg/kg was 13 and 21% of that induced by human ChG (i.p., 250 IU/rat). Meanwhile, the TP01 action was more prolonged, and 5 h after injection, the TP01-induced increase of testosterone level was 44–46% of that of human ChG [38]. It is important that TP01 was active when given orally, and at a dose of 50 mg/kg, it increased testosterone level by 230 (3 h) and 417% (5 h). The TP02 was less active when administered orally, which is probably due to reduced ability to achieve the Leydig cells as a result of rapid degradation of TP02 or its impaired absorbability in the gastrointestinal tract.

Recently, we synthesized another thienopyrimidine derivative, 5-amino-*N*-*tert*-butyl-2-(methylsulfonyl)-4-(3-(nicotinamide)phenyl)thieno[2,3-*d*]pyrimidine-6-carboxamide (TP03) and showed that TP03 at a concentration of 10^{-4} M stimulated AC activity by 213% in rat testicular membranes, and the EC_{50} value for its effect was 390 nM [39]. The TP03 when administered intraperitoneally (25 mg/kg) and orally (50 mg/kg) significantly increased the plasma testosterone level in male rats, and in this respect, it was more effective than TP01. These data indicate that TP03 can be used to develop the drugs regulating the steroidogenesis in Leydig cells.

In the case of TSHR, the most important task is the development of LMW compounds with activity of inverse agonists and neutral antagonists that inhibit the basal and hormone/antibody-stimulated activity of TSHR, respectively [40]. It was found that in Graves' disease the TSHR-stimulating antibody causes hyperstimulation of the thyroid gland, which leads to dysfunction of hypothalamic-pituitary-thyroid axis [41]. Nowadays, the effective methods for treating hyperthyroidism were not developed. The approaches that are commonly used are invasive and effective only in treating the complications of hyperthyroidism, without affecting the causes of thyroid pathology.

In 2008, Susanne Neumann and colleagues developed highly selective TSHR neutral antagonist NIDDK/CEB-5 that is structurally close to Org 43553 [42]. The treatment of TSHR-expressing HEK-EM 293 cells with 30 μM of the compound NIDDK/CEB-52 significantly decreased effects of TSH and TSHR-stimulating antibody on AC activity, and the IC_{50} value for inhibitory effect of this compound was 4.2 μM. The 24-h incubation of thyrocytes with TSH and NIDDK/CEB-52 (10 μM) led to a threefold decrease in TSH-induced expression of the gene encoding thyroid peroxidase that is required for iodination of tyrosine residues on thyroglobulin and, thereby, controls the production of thyroid hormones. In 2011, a potent TSHR inverse agonist NCGC00229600, 2-(3-((2,6-dimethylphenoxy)methyl)-4-metoxyphenyl)-3-(pyridin-3-ylmethyl)-2,3-dihydroquinazolin-4(*1H*)-one, was developed, and it, unlike NIDDK/CEB-52, suppressed both basal and stimulated TSHR activity [43]. The NCGC00229600 decreased effect of TSHR-stimulating antibody on AC activity by 30–75% in human thyrocytes and in the primary culture of fibroblasts obtained from the retro-orbital space of patients with Graves' disease [43, 44]. The ability of NCGC00229600 to suppress TSHR activity in retro-orbital fibroblasts is very important because AC overstimulation in these cells induced by TSHR-stimulating antibody is one of the main causes of ophthalmopathy, severe symptom of Graves' disease.

Newly developed neutral furan-containing antagonist NCGC00242364 selectively inhibited TSH-induced AC activity in TSHR-expressing cells and suppressed the effects of TSH and TSHR-stimulating antibody on thyroxine production and expression of the genes encoding Na^{+}-I^{-} cotransporter and thyroid peroxidase in female mice BALB/c [40]. These data demonstrate high efficiency of the NCGC00242364 and its analogs in the treatment of Graves' disease. It is very important that these compounds had no effect on the basal TSHR activity, which allows avoiding the development of hypothyroid states typical for TSHR inverse agonists [40]. The inverse agonists are more suitable to treat nonautoimmune hyperthyroidism caused by activating mutations in TSHR.

The development of LMW ligands of LH/ChG-R and TSHR allows creating a wide range of drugs for selective regulation of hypothalamic-pituitary-gonad and hypothalamic-pituitary-thyroid axes. The LMW agonists of LH/ChG-R can be used to induce ovulation and fertilization in females and to enhance the production of steroid hormones in males with androgen deficiency and hypogonadotropic states. The inverse agonists and neutral antagonists are useful for contraception and treatment of hormone-dependent tumors of the reproductive system. The LMW antagonists and inverse agonists are the promising drugs for prevention of hyperactivation of the thyroid gland by TSH and TSHR-stimulating antibody, while the full TSHR agonists can be used in the diagnostics of thyroid cancer and in radioiodine therapy. It should be noted that LMW regulators are active when administered orally because they do not degrade in the gastrointestinal tract and are easily absorbed into blood. They are effective even when they act on mutant forms of GPCRs that are not capable to processing and translocation into the plasma membrane. The LMW agonists act as highly selective pharmacological chaperones for the cognate receptors. The agonists penetrate into the cells, bind to mutant

GPCRs located in the cytoplasm or endoplasmic reticulum, stabilize their active conformation, preventing receptor degradation in proteasomes and facilitating GPCR translocation into the plasma membrane [45, 46].

Acknowledgements

This work was supported by the Russian Foundation of Basic Investigations (project No 16-04-00126) and by the state assignment of FASO of Russia.

Author details

Alexander O. Shpakov and Kira Viktorovna Derkach*

*Address all correspondence to: derkatch_k@list.ru

I.M. Sechenov institute of Evolutionary Physiology and Biochemistry RAS, St. Petersburg, Russia

References

[1] Luttrell LM. Transmembrane signaling by G protein-coupled receptors. Methods in Molecular Biology. 2006;**332**:3-49. DOI: 10.1385/1-59745-048-0:1

[2] Liapakis G, Cordomí A, Pardo L. The G-protein coupled receptor family: Actors with many faces. Current Pharmaceutical Design. 2012;**18**:175-185

[3] Strotmann R, Schrock K, Boselt I, Staubert C, Russ A, Schoneberg T. Evolution of GPCR: Change and continuity. Molecular and Cellular Endocrinology. 2011;**331**:170-178. DOI: 10.1016/j.mce.2010.07.012

[4] Overington JP, Al-Lazikani B, Hopkins AL. How many drug targets are there? Nature Reviews. Drug Discovery. 2006;**5**:993-996. DOI: 10.1038/nrd2199

[5] Shoichet BK, Kobilka BK. Structure-based drug screening for G-protein-coupled receptors. Trends in Pharmacological Sciences. 2012;**33**:268-272. DOI: 10.1016/j.tips.2012.03.007

[6] Covic L, Gresser AL, Talavera J, Swift S, Kuliopulos A. Activation and inhibition of G protein-coupled receptors by cell-penetrating membrane-tethered peptides. Proceedings of the National Academy of Sciences of the United States of America. 2002;**99**:643-648. DOI: 10.1073/pnas.022460899

[7] Shpakov AO. Signal protein-derived peptides as functional probes and regulators of intracellular signaling. Journal of Amino Acids. 2011;**2011**:656051. DOI: 10.4061/2011/656051

[8] Tressel SL, Koukos G, Tchernychev B, Jacques SL, Covic L, Kuliopulos A. Pharmacology, biodistribution, and efficacy of GPCR-based pepducins in disease models. Methods in Molecular Biology. 2011;**683**:259-275. DOI: 10.1007/978-1-60761-919-2_19

[9] Shpakov AO, Shpakova EA. The use of peptides derived from G protein-coupled receptors and heterotrimeric G proteins in the study of their structure and functions. Chapter 4. In: Protein Purification and Analysis III—Methods and Applications. 2014. iConcept Press Ltd. ISBN: 978-1-922227-65-2

[10] Granier S, Terrillon S, Pascal R, Demene H, Bouvier M, et al. A cyclic peptide mimicking the third intracellular loop of the V_2 vasopressin receptor inhibits signaling through its interaction with receptor dimer and G protein. The Journal of Biological Chemistry. 2004;**279**:50904-50914. DOI: 10.1074/jbc.M405089200

[11] Derkach KV, Shpakova EA, Titov AM, Shpakov AO. Intranasal and intramuscular administration of lysine-palmitoylated peptide 612-627 of thyroid-stimulating hormone receptor increases the level of thyroid hormones in rats. International Journal of Peptide Research and Therapeutics. 2015;**21**:249-260. DOI: 10.1007/s10989-014-9452-6

[12] Agarwal A, Tressel SL, Kaimal R, Balla M, Lam FH, Covic L, Kuliopulos A. Identification of a metalloprotease-chemokine signaling system in the ovarian cancer microenvironment: Implications for antiangiogenic therapy. Cancer Research. 2010;**70**:5880-5890. DOI: 10.1158/0008-5472.CAN-09-4341

[13] O'Callaghan K, Lee L, Nguyen N, Hsieh MY, Kaneider NC, Klein AK, Sprague K, Van Etten RA, Kuliopulos A, Covic L. Targeting CXCR4 with cell-penetrating pepducins in lymphoma and lymphocytic leukemia. Blood. 2012;**119**:1717-1725. DOI: 10.1182/blood-2011-04-347518

[14] Shpakov AO, Gur'yanov IA, Kuznetsova LA, Plesneva SA, Shpakova EA, Vlasov GP, Pertseva MN. Studies of the molecular mechanisms of action of relaxin on the adenylyl cyclase signaling system using synthetic peptides derived from the LGR7 relaxin receptor. Neuroscience and Behavioral Physiology. 2007;**37**:705-714. DOI: 10.1007/s11055-007-0071-y

[15] Shpakov AO, Shpakova EA, Tarasenko II, Derkach KV, Vlasov GP. The peptides mimicking the third intracellular loop of 5-hydroxytryptamine receptors of the types 1B and 6 selectively activate G proteins and receptor-specifically inhibit serotonin signaling via the adenylyl cyclase system. International Journal of Peptide Research and Therapeutics. 2010;**16**:95-105. DOI: 10.1007/s10989-9208-x

[16] Shpakov AO, Shpakova EA, Derkach KV. The sensitivity of the adenylyl cyclase system in rat thyroidal and extrathyroidal tissues to peptides corresponding to the third intracellular loop of thyroid-stimulating hormone receptor. Current Topics in Peptide & Protein Research. 2012;**13**:61-73

[17] Shpakov AO, Shpakova EA, Tarasenko II, Derkach KV. Receptor and tissue specificity of the effect of peptides corresponding to intracellular regions of the serpentine type

receptors. Biochemistry (Moscow) Supplement Series A Membrane and Cell Biology. 2012;**64**:16-25. DOI: 10.1134/S1990747811060146

[18] Forsman H, Bylund J, Oprea TI, Karlsson A, Boulay F, Rabiet MJ, Dahlgren C. The leukocyte chemotactic receptor FPR2, but not the closely related FPR1, is sensitive to cell-penetrating pepducins with amino acid sequences descending from the third intracellular receptor loop. Biochimica et Biophysica Acta. 2013;**1833**:1914-1923. DOI: 10.1016/j.bbamcr.2013.03.026

[19] Zhang P, Gruber A, Kasuda S, Kimmelstiel C, O'Callaghan K, Cox DH, Bohm A, Baleja JD, Covic L, Kuliopulos A. Suppression of arterial thrombosis without affecting hemostatic parameters with a cell-penetrating PAR1 pepducin. Circulation. 2012;**126**:83-91. DOI: 10.1161/CIRCULATIONAHA.112.091918

[20] Cisowski J, O'Callaghan K, Kuliopulos A, Yang J, Nguyen N, Deng Q, Yang E, Fogel M, Tressel S, Foley C, Agarwal A, Hunt SW 3rd, McMurry T, Brinckerhoff L, Covic L. Targeting protease-activated receptor-1 with cell-penetrating pepducins in lung cancer. The American Journal of Pathology. 2011;**179**:513-523. DOI: 10.1016/j.ajpath.2011.03.025

[21] Foley CJ, Luo C, O'Callaghan K, Hinds PW, Covic L, Kuliopulos A. Matrix metalloprotease-1a promotes tumorigenesis and metastasis. The Journal of Biological Chemistry. 2012;**287**:24330-24338. DOI: 10.1074/jbc.M112.356303

[22] Michael ES, Kuliopulos A, Covic L, Steer ML, Perides G. Pharmacological inhibition of PAR2 with the pepducin P2pal-18S protects mice against acute experimental biliary pancreatitis. American Journal of Physiology. Gastrointestinal and Liver Physiology. 2013;**304**:G516-G526. DOI: 10.1152/ajpgi.00296.2012

[23] Shpakov AO. Participation of charged amino acid residues of cytoplasmic loops of serpentine type receptors in the process of transmission of hormonal signal. Journal of Evolutionary Biochemistry and Physiology. 2003;**39**:266-280. DOI: 10.1023/A:1026135522916

[24] Shpakov AO, Shpakova EA, Derkach KV. A comparative study of biological activity of peptides, the derivatives of the third intracellular loop of rat luteinizing hormone receptor, modified by hydrophobic radicals at the N- and C-termini. Current Topics in Peptide & Protein Research. 2013;**14**:1-12

[25] Puett D, Li Y, DeMars G, Angelova K, Fanelli F. A functional transmembrane complex: The luteinizing hormone receptor with bound ligand and G protein. Molecular and Cellular Endocrinology. 2007;**260**:126-136. DOI: 10.1016/j.mce.2006.05.009

[26] Kleinau G, Krause G. Thyrotropin and homologous glycoprotein hormone receptors: Structural and functional aspects of extracellular signaling mechanisms. Endocrine Reviews. 2009;**30**:133-151. DOI: 10.1210/er.2008-0044

[27] Troppmann B, Kleinau G, Krause G, Gromoll J. Structural and functional plasticity of the luteinizing hormone/choriogonadotrophin receptor. Human Reproduction Update. 2013;**19**:583-602. DOI: 10.1093/humupd/dmt023

[28] van Straten NC, Schoonus-Gerritsma GG, van Someren RG, Draaijer J, Adang AE, Timmers CM, Hanssen RG, van Boeckel CA. The first orally active low molecular weight agonists for the LH receptor: Thienopyr(im)idines with therapeutic potential for ovulation induction. Chembiochem. 2002;**3**:1023-1026. DOI: 10.1002/1439-7633(20021004)3:10<1023::AID-CBIC1023>3.0.CO;2-9

[29] Shpakov AO, Shpakova EA. Low-molecular regulators of polypeptide hormone receptors containing LGR-repeats. Biochemistry (Moscow) Supplement Series B: Biomedical Chemistry. 2009;**3**:351-360. DOI: 10.1134/S1990750809040040

[30] Lane JR, IJzerman AP. Allosteric approaches to GPCR drug discovery. Drug Discovery Today: Technologies. 2013;**10**:219-221. DOI: 10.1016/j.ddtec.2013.01.006

[31] van Koppen CJ, Zaman GJ, Timmers CM, Kelder J, Mosselman S, van de Lagemaat R, Smit MJ, Hanssen RG. A signaling-selective, nanomolar potent allosteric low molecular weight agonist for the human luteinizing hormone receptor. Naunyn-Schmiedeberg's Archives of Pharmacology. 2008;**378**:503-514. DOI: 10.1007/s00210-008-0318-3

[32] Gilchrist RL, Ryu KS, Ji I, Ji TH. The luteinizing hormone/chorionic gonadotropin receptor has distinct transmembrane conductors for cAMP and inositol phosphate signals. The Journal of Biological Chemistry. 1996;**271**:19283-19287. DOI: 10.1074/jbc.271.32.19283

[33] van de Lagemaat R, Timmers CM, Kelder J, van Koppen C, Mosselman S, Hanssen RG. Induction of ovulation by a potent, orally active, low molecular weight agonist (Org 43553) of the luteinizing hormone receptor. Human Reproduction. 2009;**24**:640-648. DOI: 10.1093/humrep/den412

[34] van de Lagemaat R, Raafs BC, van Koppen C, Timmers CM, Mulders SM, Hanssen RG. Prevention of the onset of ovarian hyperstimulation syndrome (OHSS) in the rat after ovulation induction with a low molecular weight agonist of the LH receptor compared with hCG and rec-LH. Endocrinology. 2011;**152**:4350-4357. DOI: 10.1210/en.2011-1077

[35] Gerrits M, Mannaerts B, Kramer H, Addo S, Hanssen R. First evidence of ovulation induced by oral LH agonists in healthy female volunteers of reproductive age. The Journal of Clinical Endocrinology and Metabolism. 2013;**98**:1558-1566. DOI: 10.1210/jc.2012-3404

[36] Shpakov AO, Dar'in DV, Derkach KV, Lobanov PS. The stimulating influence of thienopyrimidine compounds on the adenylyl cyclase systems in the rat testes. Doklady. Biochemistry and Biophysics. 2014;**456**:104-107. DOI: 10.1134/S1607672914030065

[37] Shpakov AO, Derkach KV, Dar'in DV, Lobanov PS. Activation of adenylyl cyclase by thienopyrimidine derivatives in rat testes and ovaries. Cell and Tissue Biology. 2014;**8**:400-406. DOI: 10.1134/S1990519X14050071

[38] Derkach KV, Dar'in DV, Lobanov PS, Shpakov AO. Intratesticular, intraperitoneal, and oral administration of thienopyrimidine derivatives increases the testosterone level in male rats. Doklady Biological Sciences. 2014;**459**:326-329. DOI: 10.1134/S0012496614060040

[39] Derkach KV, Legkodukh AS, Dar'in DV, Shpakov AO. The stimulating effect of thienopyrimiduines, the analogues of Org 43553, on adenylyl cyclase activity in the testes and on testosterone production in male rats. Tsitologiia. 2016;**58**:602-609 Russian

[40] Neumann S, Nir EA, Eliseeva E, Huang W, Marugan J, Xiao J, Dulcey AE, Gershengorn MC. A selective TSH receptor antagonist inhibits stimulation of thyroid function in female mice. Endocrinology. 2014;**155**:310-314. DOI: 10.1210/en.2013-1835

[41] Rapoport B, McLachlan SM. The thyrotropin receptor in graves' disease. Thyroid. 2007; **17**:911-922. DOI: 10.1089/thy.2007.0170

[42] Neumann S, Kleinau G, Costanzi S, Moore S, Jiang JK, Raaka BM, Thomas CJ, Krause G, Gershengorn MC. A low-molecular-weight antagonist for the human thyrotropin receptor with therapeutic potential for hyperthyroidism. Endocrinology. 2008;**149**:5945-5950. DOI: 10.1210/en.2008-0836

[43] Neumann S, Eliseeva E, McCoy JG, Napolitano G, Giuliani C, Monaco F, Huang W, Gershengorn MC. A new small-molecule antagonist inhibits Graves' disease antibody activation of the TSH receptor. The Journal of Clinical Endocrinology and Metabolism. 2011;**96**:548-554. DOI: 10.1210/jc.2010-1935

[44] Neumann S, Pope A, Geras-Raaka E, Raaka BM, Bahn RS, Gershengorn MC. A drug-like antagonist inhibits thyrotropin receptor-mediated stimulation of cAMP production in Graves' orbital fibroblasts. Thyroid. 2012;**22**:839-843. DOI: 10.1089/thy.2011.0520

[45] Newton CL, Whay AM, McArdle CA, Zhang M, van Koppen CJ, van de Lagemaat R, Segaloff DL, Millar RP. Rescue of expression and signaling of human luteinizing hormone G protein-coupled receptor mutants with an allosterically binding small-molecule agonist. Proceedings of the National Academy of Sciences of the United States of America. 2011;**108**:7172-7176. DOI: 10.1073/pnas.1015723108

[46] Shpakov AO. New achievements in the development and study of the mechanisms of action of the low molecular weight agonists of receptors of the thyroid-stimulating and the luteinizing hormones. Tsitologiia. 2015;**57**:167-176. PMID: 26021165. Russian

Chapter 7

Ontogenetic Development of Neurophysiological Mechanisms Underlying Language Processing

Diana M. Guillemard (Tsaparina),
Michail N. Tsitseroshin, Alexandr N. Shepovalnikov,
Elizaveta I. Galperina, Ekaterina A. Panasevich,
Ekaterina E. Kats, Larisa G. Zaytseva and
Olga V. Kruchinina

Additional information is available at the end of the chapter

http://dx.doi.org/10.5772/intechopen.73654

Abstract

During the last 20 years, new data on the neurophysiological mechanisms underlying different types of cognitive activity, especially speech and its ontogenetic formation, were obtained in the Laboratory of Children's Neurophysiology headed by Prof. M.N. Tsitseroshin. Using the analysis of the spatial-temporal structure of regional interactions of cortical bioelectric potentials (so-called functional connectivity), we investigated how specific language levels, such as phonology, grammar, and semantics, are represented in the brain. The data obtained in children vs. adults indicate that the speech perception and production require joint and extremely coordinated activities of both hemispheres, along with the obligatory and differentiated involvement of "classic" speech centers in the left hemisphere, especially Wernicke's area. Another line of our research is to explore the differences, which arise during verbal processing in adults and children with impaired vs. non-impaired speech, particularly with alalia, dysarthria and stuttering, using behavioral and EEG data. Our data obtained in children vs. adults allow assessing the degree of maturity in the organization of the central processes of maintaining the studied types of verbal activity in children of different ages. These data allow expanding modern concepts about the brain mechanisms of verbal activity in children in the norm and pathology.

Keywords: speech processing, phonology, grammar, semantics, EEG, brain, development, interhemispheric interactions, speech pathology

1. Introduction

A study of the functional interaction between various parts of the cerebral cortex (so-called functional connectivity) is of fundamental importance for understanding the neural basis of cognition [1–14]. A growing number of studies that have appeared in the last decade testify the renewed interest in functional interactions among different brain parts, in both the background state and under cognitive activity [13, 15]. Unraveling functional connections in the human brain, with a main focus on the formation of the integrative brain mechanisms, has been a long-term research goal of the Laboratory of Child's Neurophysiology at the Sechenov Institute of Evolutionary Physiology and Biochemistry of the Russian Academy of Sciences headed by Prof. M.N. Tsitseroshin. The original methods designed in the laboratory to analyze the primary and secondary EEG parameters provided new insight into the genesis of these mechanisms. It was shown that the spatial-temporal structure of interregional interactions of cortical bioelectrical activity (i.e., functional connectome) in healthy subjects is highly spatially ordered (**Figure 1**).

It was shown for the first time that the structure of the distant EEG statistical relationships in adults does not arise randomly or chaotically, but occurs in a strictly ordered manner (**Figure 1A**). The implementation of various types of cognitive activity is associated only with transitional changes in the spatial structure of the interregional interactions while maintaining the integration processes of the brain activity characteristic of the resting state. Our studies have shown that this ordered structure of cortical interactions retains its topological invariance also in other background states of the brain, in particular, at different stages of natural and hypnotic sleep. Stable distortion of the ordered structure of intracortical interactions was detected only in case of cerebral pathology [5, 16–19] (**Figure 1B**).

During brain maturation, this order gradually increases, reflecting the formation of integrative relationships in the cortex and subcortical structures. This process creates a stable morphofunctional basis for the effective performance of learning processes and optimization of adaptive reactions. The authors consider this phenomenon as one of the fundamental features of the whole brain's functional organization required for the brain's normal activity. The orderliness of the spatial organization of brain biopotential field reflects the functioning of main integrative mechanisms of the brain. Based on the data obtained using factor analysis of EEG, it is possible to specify some morphofunctional brain systems (factors), which are directly involved in the processes of brain activity integration. According to these ideas, the factor I reflects general properties of the wave bioelectric processes occurring on the convexital surface of the cortex. It also reflects a degree of generalized nonspecific modulating influences of the brainstem and medulla on the cortical fields' activity. Factor II reflects the frontal-occipital relationships, that is, processes of activity integration for the anterior and posterior portions of each hemisphere by means of long association pathways and thalamocortical associative systems. Factor III reflects interhemispheric interactions, that is, processes providing coupled activity of the cerebral hemispheres by means of the commissural fibers with the involvement of the striopallidar and limbic systems.

http://dx.doi.org/10.5772/intechopen.73654

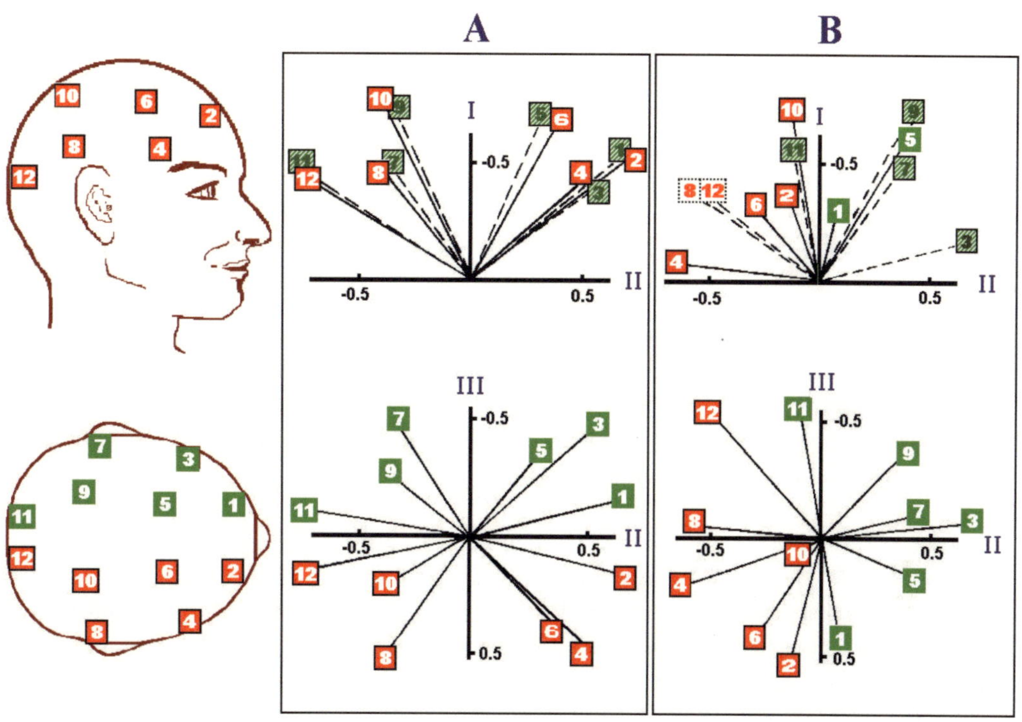

Figure 1. New method of EEG data representation in 3D factor space. (A) In the norm, the location of the radius vectors (the angles between vectors reflect the cross-correlation between EEG processes in different cortical areas) and electrodes is very similar, indicating strictly ordered spatial interactions of cortical areas' activity. (B) In case of cerebral pathology, the revealed topological violations of the radius vectors' location. At the top—projections of EEG radius vectors on the plane of factors I–II. At the bottom—same on the plane of factors II–III. Numbers of EEG radius vectors (1–12) that correspond to EEG leads in the left charts.

Methodological approaches used in our laboratory, including the priority methods (patent RU № 2177716 of 10.01.2002), allow quantitative assessment of the contribution of the left and right hemispheric cortical fields to the organization of neurophysiological mechanisms underlying various types of cognitive activity, including speech [7].

2. Neurophysiological mechanisms underlying language development

Current data on the principles of organization of the central nervous system and mental activities allow linking the processes of cognitive activity with the simultaneous involvement of numerous spatially distributed cortical and subcortical structures. An appropriate way to study patterns of cortical interactions is analysis of the spatiotemporal organization of brain activity oscillations. Currently, it is of no doubt that language processing relies on a widespread network of brain regions [1–5, 20–26]. Great importance is also attached to the role of cortical networks and functional connections between different parts of the brain, in both adults and children [21, 22, 27–33].

Despite a long period of study, the question about a special role of the left and right hemispheres in providing the speech function as well as on the degree of involvement of interhemispheric relationships in these processes is still at the peak of interest [34]. Considering the well-known concepts of a greater significance of the right hemisphere for the speech processes in children vs. adults [35, 36], a study of the central mechanisms of the speech function formation in the children's ontogeny is of particular relevance.

The issues of the levels of involvement of the left and right hemispheres in supporting verbal activity are of continuing interest of investigators; however, their analysis sometimes yields mutually contradictory results. The use of the state-of-the-art neuroimaging methods, such as functional magnetic resonance imaging (fMRI), has unexpectedly increased the number of reports with the evidence of predominantly left-sided location of the neural centers responsible for speech perception and production [32, 37–39]. Thus, Binder et al. [37], while reviewing 120 articles on cortical locations of the areas associated with semantic speech processes, reported that 68% of 1145 fMRI or PET activations were detected in the left hemisphere cortex areas, 32% in the right hemisphere, and 10% in the cerebellum. The locations of right-sided foci were generally homologous to those of several left-sided foci—in the angular gyrus, posterior middle temporal gyrus, and cingulate gyrus. But, these data summation by a special algorithm led the authors to conclude mainly left-sided lateralization for verbal semantic processes. Price [38] has come to similar conclusions considering 100 fMRI speech studies reported in 2009. However, other authors, despite some left hemisphere advantages, defend the bilateral cerebral organization of language processing. This viewpoint was emphasized in the Hickock and Poeppel model [20]. The studies by Hagman et al. [40], using fMRI combined with diffusion tractography, pointed out that the subject gender has a significant influence on the speech processing lateralization. The results of spectral coherence and cross-correlation analysis of EEG-data provided a highly reproducible result associated with significant changes in both hemispheres during verbal processing, with increased levels of hemispheric interactions [21, 22, 30, 31]. Two earlier studies in this area [41, 42] emphasized that changes in EEG spectral power and coherence during speech processing are seen in both hemispheres showing no signs of any predominant lateralization ("These results do not indicate any lateralized EEG changes during verbal tasks") ([42], p. 357). Our studies of cortex regional interaction organization in children and adults, during different verbal tasks, have been shown the long-distance connection activation between cortical biopotentials of the left and right hemispheres [1–5, 7, 43, 44]. According to Ross [24], "functional localization in the brain is a robust dynamic and four-dimensional process. It is a learned phenomenon driven over time by large-scale, spatially distributed neural networks seeking to efficiently maximize processing, storage, and manipulation of information for cognitive and behavioral operations." Yourganov et al. [45] also pointed out that connectome-based approaches should be used in combination with lesion-based ones to fully elucidate whether structurally damaged or disconnected regions relate to the aphasic impairment and its recovery.

During the past 20 years, new data have been obtained in our laboratory on neurophysiological mechanisms underlying different types of cognitive activity, especially speech and its formation in ontogenesis. Special attention was paid to analysis of interhemispheric interactions and the involvement of the left and right hemispheres in speech. Our investigations were focused on the central mechanisms underlying speech processing at different language levels: phonological, semantic, syntactical, and lexical. On the other hand, we investigated both analytical speech processing (auditory perception and analysis of verbal signals) and synthetic speech processing (mental formation of words from sets of phonemes, mental formation of sentences from sets of words). We also investigated EEG correlates of verbal task performance in healthy adults and children as well as children with dysarthria, alalia, and typical speech development. Our results obtained in children vs. adults allow us to assess the degree of maturity of the central processes providing the maintenance of the studied types of verbal activity in children at different ages. The results of our research allow expanding modern concepts about the brain mechanisms of verbal activity in children in the norm and pathology; they can also assist in developing effective approaches to early correction of speech disorders.

At the semantic and syntactical language levels, we investigated the spatial organization of brain biopotentials in healthy adults during the perception of sentences with syntactical or semantic mistakes. At the syntactical language level, we used sentences containing the following syntactical mistakes: (1) gender agreement errors: in personal endings of verbs, for example—"Pod stolom sidel sobaka"("A dog [feminine] sat [masculine] under the table") and in gender endings of adjectives—"Devochka nadela goluboe yubku ("The girl put on a blue [neuter] skirt [feminine]"); (2) mistakes in Russian nouns: in inflections of nouns—"Ya prochital interesnuyu knizhkoy" ("I have read an interesting [accusative case] book [instrumental case]"). As for semantic, the stimulus material contained the following types of semantic mistakes: (1) mistakes in constructions reflecting the time sequence of events ("Winter comes after spring") and (2) constructions containing a semantic paradox ("A tortoise overtook a deer"). Our results revealed changes in distant biopotential relationships, which were both similar and specific to the type of mistakes. In both cases, the interhemispheric biopotential relationships increased significantly as compared to eyes-closed resting conditions. At the same time, almost the total absence of changes in intrahemispheric EEG correlations was observed. An increase in contralateral interactions was particularly characteristic of the activity of Broca and Wernicke's areas and the middle temporal zone of the right hemisphere. Distant EEG relationships in Broca's area increased more significantly during the detection of semantic rather than grammar mistakes in adults. Obtained data also showed that detection of grammatical and syntactic mistakes was supported by the intercentral connections related to the functional system responsible for the detection not only of verbal but also any other stimuli [3] (**Figure 2**).

The recent data [46] provide evidence that sensitivity to syntactic complexity is widespread across the language system, contrary to many previous neuroimaging studies that reported only a few, localized foci of syntactic complexity effects.

The other line of our research addresses the neurophysiological mechanisms underlying the maintenance of mental speech activity associated with the synthesis of speech units from elementary components (mental composing words from a set of phonemes and composing sentences

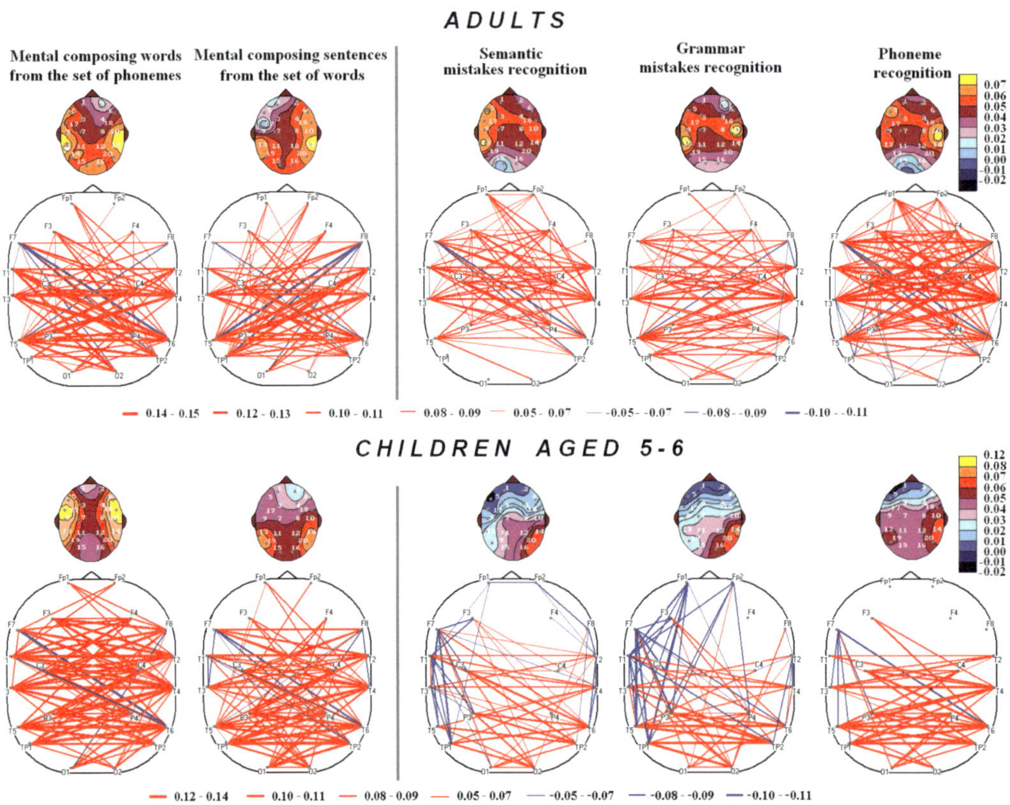

Figure 2. Increase in interhemispheric interactions during speech perception and production in adults and children aged 5–6 years. Left side—mental synthesis of speech units: composition of words from the set of auditory presented phonemes and composition of sentences from a set of auditory presented words. Right side—detection of semantic and syntactic mistakes in auditory presented sentences; analysis of verbal material during phoneme recognition in auditory presented words [4].

from a set of words) in healthy adults [2] and children aged 5–6 years [4]. During the mental composition of words from an auditory presented set of phonemes (phonemic synthesis) or sentences from a set of words, the adults exhibited specific changes in the spatial structure of the statistical EEG relationships with a significant increase in the interhemispheric interactions. During the performance of both tasks, changes in the interhemispheric interactions were typical for the temporal, temporo-parieto-occipital (TPO), inferofrontal, and occipital areas of both hemispheres (**Figure 2**). Phonemic synthesis was associated with a more pronounced increase in the contralateral interactions in the left hemisphere, as well as with generating sentences from words in the right hemisphere. The coherence analysis of EEG showed maximal changes in the delta, theta, and beta frequency bands with rather slight changes in the alpha frequency band (**Figure 3**, adults). For all frequency bands, changes in EEG coherence were the greatest in Wernicke's and the TPO areas of the right and left hemispheres during the performance of both tasks, especially during the phonemic synthesis. These findings suggest that neurophysiological processes underlying mental generation of words and sentences require coordinated activity of the left and right hemispheres, which is accompanied by an increase in the interhemispheric interactions in EEG, especially in the temporal, inferofrontal, and TPO areas.

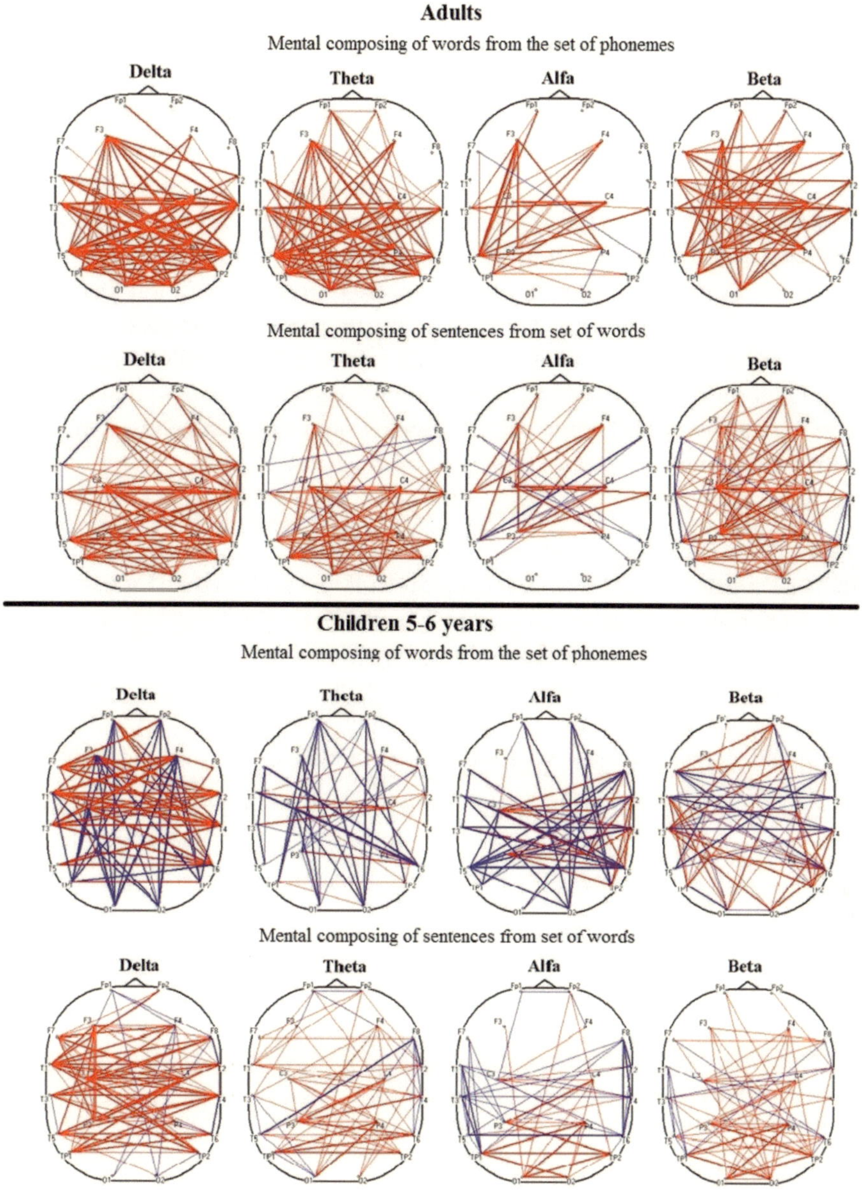

Figure 3. Changes in the spatial structure of EEG coherence connections in different frequency bands (delta, theta, alpha, beta) in adults while composing words from a set of phonemes and sentences from a set of words (baseline compared). Below—the same in children aged 5–6 years. Red lines show increases and blue lines show decreases of EEG coherence connections. Significant changes at $p \leq 0.05$ are shown.

In children aged 5–6 years, the composition of words from a set of auditory presented phonemes and of sentences from sets of words showed significant changes in the spatial structure of interregional EEG relationships [4] (**Figure 2**). Both tasks evoked significant increases in interhemispheric interactions between different cortical areas. Cross-correlation relationships showed a particular increase in biopotential oscillations in the inferior frontal areas of both hemispheres, as well as in the anterior, intermediate, and posterior temporal areas, central and parietal areas of the cortex, and TPO areas of the left and right hemispheres. However, some differences in the topical features of changes in EEG cross-correlation relationships were also observed (**Figure 2**). In children, maximal changes toward increased interhemispheric connections were observed for EEG connections between the frontal areas (particularly inferior and posterior) and all temporal cortical of the right and left hemispheres (i.e., anterior, intermediate, and posterior) and TPO areas during word composition task. In turn, maximal increases while composing sentences from a set of words were characteristic of interhemispheric EEG connections in the posterior cortical areas (including the posterior temporal ones), TPO areas, and occipital areas of both hemispheres. It should be noted that in children, the activity associated with the formation of words represents a more difficult task than that associated with the formation of sentences. The rate of composing words from sounds was much slower in children than the rate of composing sentences from words. This may be due to the children preschool age, and they are still unable to read and write, so they do not operate with visual images while inventing words, comparing to adults [2].

This verbal task seems to be more abstract and difficult for children than for adults. Some EEG coherence interactions decrease (especially long distance) in children while composing words may be due to a greater subjective difficulty of the task being performed. Weiss and Mueller [22] pointed that successful creative speech activity is accompanied by increases in coherence between EEG signals from distant cortical zones while increased subjective difficulty in performing linguistic tasks correlates with a decrease of EEG coherence connections.

In children, some differences in theta and delta frequency bands were observed: mainly a decrease in coherent interactions during composing words and, conversely, an increase in the coherent interactions of EEG oscillations during composing sentences (**Figure 3**, children).

Bastiaansen et al. also demonstrated an increase in spectral power of the theta band during performing sentence-processing tasks [47]. Since the increase in theta activity during verbal tasks is associated with memory processes, it is likely that increase in the coherence of EEG oscillations in the theta range can be due to features of sentence composing associated with an increase in the efficiency of information processes [21, 22].

Thus, in children of preschool age, composing of speech units from more elementary components revealed complex patterns of changes in the distant interactions of bioelectrical activity in different areas of both hemispheres. The complexity of these patterns with the decrease and increase in the correlation and coherent EEG interactions seems to reflect the formation of the spatial structure of systemic interactions between these cortical areas, whose joint action is necessary to support verbal utterance generation during the postnatal ontogeny period.

To explore the regional brain interactions biopotentials at the phonological language level, we compared subject's performance results of three phonological tasks: phoneme recognition in words, composing a word from a set of phonemes, and controlled oral word association test. Cross-correlation and coherent analyses of EEG revealed differences in the spatial-temporal structure of brain biopotential interaction in adults. A significant increase of the interhemispheric relations was observed during phoneme recognition, especially in theta and delta frequency bands between temporal areas of both hemispheres [44] (**Figure 2**). During word composition task, changes in the hemispheric interactions were most pronounced in the temporal, especially Wernike's and temporo-parieto-occipital areas of both hemispheres [2] (**Figure 2**). A remarkable increase in regional interactions in the occipital areas accompanied by a decrease in interactions in frontal areas was observed during controlled word association. Our data showed that differences in the structure of regional interactions of brain biopotentials during verbal tasks at the phonemic language level depend on different task performance in spite of phonological specificity tasks.

The essential role of joint activity of both cerebral hemispheres in providing speech was also revealed during the analysis of the violations of systemic brain activity organization in children with developmental speech disorders such as alalia, dysarthria, and stuttering [1, 5, 6, 16]. A study of irregularities in the brain activity systemic organization, in children with various types of speech pathology, also allows deeper understanding of processes of formation of the neurophysiological mechanisms underlying language functions. In 12 children aged 3–4 years with motor alalia and 15 children aged 5–6 years with dysarthria were detected both significant differences in the nature and localization of violations of the systemic interaction of brain potentials and were revealed indicators of similarity (**Figure 4**) [16].

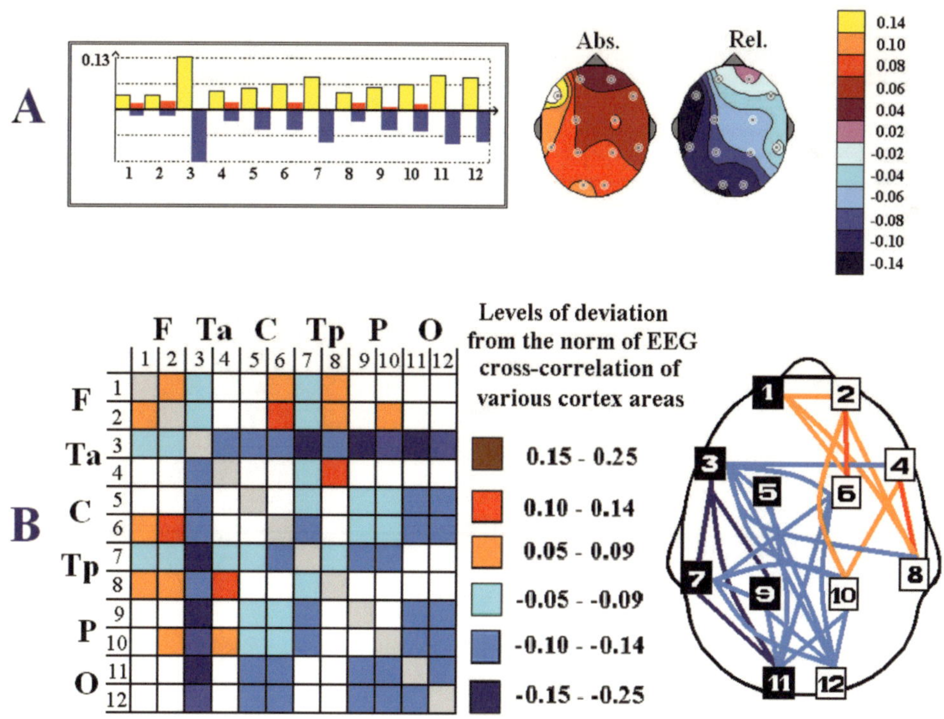

Figure 4. Disturbances in interregional interaction of cortical fields' activity in 3-year-old children with motor alalia.

To specify the role of hemispheric interactions in pathogenesis of these verbal impairments, it is important to consider the data showing that in case of both alalia and dysarthria there were revealed two types of changes in systemic interaction of brain cortical biopotentials with the direction of these changes differing in the right and left hemispheres [1]. In all children with these types of verbal pathology, against the background of a decrease in EEG connections of one of the hemispheres, the interregional interaction of bioelectrical activity in the brain cortex also significantly exceeded the interaction in the contralateral hemisphere, as compared to the normative level; this was especially characteristic of children with dysarthria (**Figure 5**).

It cannot be ruled out that the revealed exceeding over the normal level of regional interaction of the cortical bioelectrical activity in children with alalia and dysarthria can be pathological, making the adequate development of verbal functions difficult.

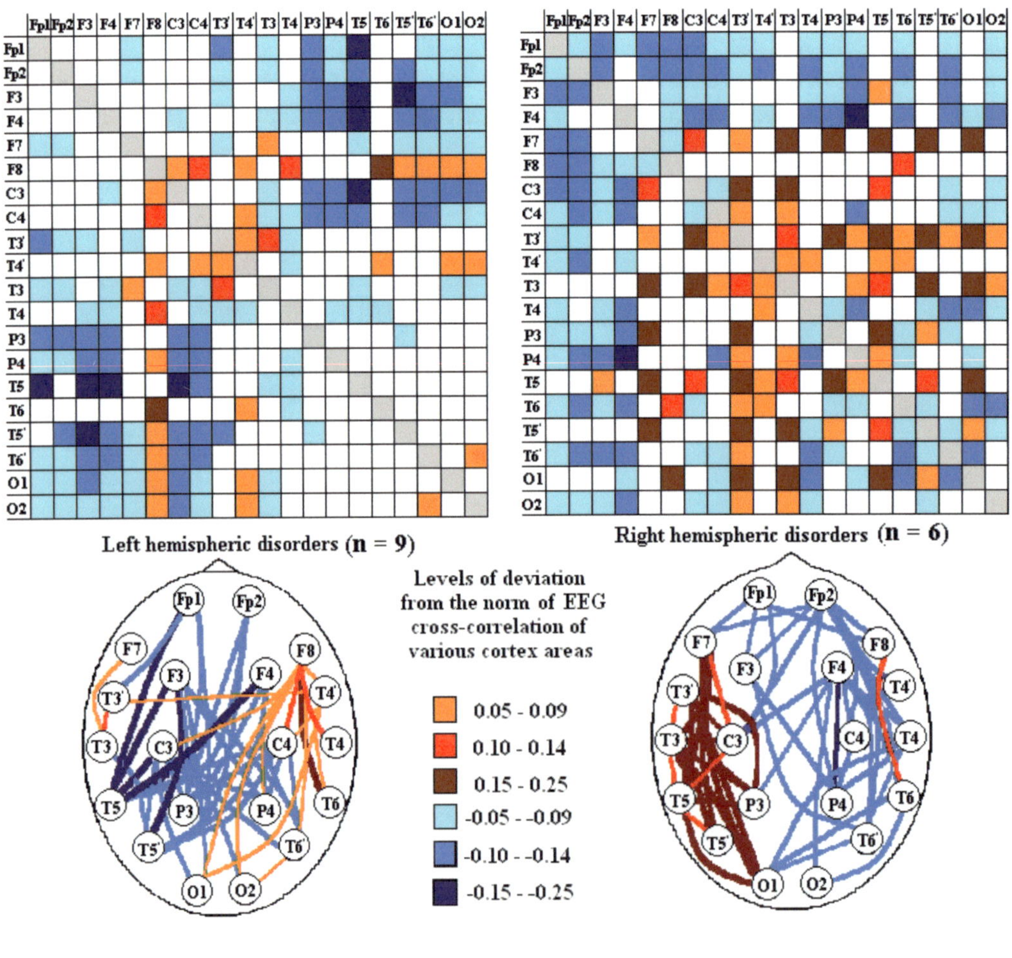

Figure 5. Infringements of systemic interaction of brain biopotentials in children aged 5–6 years with dysarthria.

Thus, the evidence of the essential role of joint activity of both brain hemispheres in providing speech was obtained by the analysis of the violations of the systemic organization of brain activity in children with deviations of speech development, such as alalia, dysarthria, and stuttering.

The other series of our studies [48, 49] addresses the role of multiple factors that determine the individual variety of neurophysiological mechanisms providing learning and using two languages. Speech formation is influenced by both common factors of bilingual and monolingual and bilingual special factors. Common factors include genetic and environmental effects explaining the individual differences in the morphofunctional organization of the speech function. Apparently, bilinguals have a large variability of the central mechanisms providing speech function due to various factors affecting language environment. These conditions include the age of the second-language acquisition, language proficiency, linguistic similarity of languages, the method of language acquisition, and the intensity and situation of use, and individual sensitivity to the above factors can be different. Unique features of the brain, existing only at the initial stages of postnatal ontogenesis, need for language acquisition, are effectively used in the situation of bilingualism. With age, the individual variability of EEG activation patterns during verbal activity in bilinguals increases. Perhaps with age during second-language acquisition, the brain is forced to manipulate a large number of additional cerebral mechanisms responsible for speech functions. There is a serious reason to believe that second-language acquisition contributes to the expansion of the functional capabilities of the brain and creates the basis for successful cognitive activity.

3. Conclusions

The results of long-term investigations conducted in our laboratory were presented in the collective monograph "The Formation of the Integrative Functions of the Brain" [7], edited by N.P. Bechtereva. According to the pivotal idea running through this monograph, the important feature of the integrative mechanisms of the brain consists in providing at the consecutive stages of postnatal ontogenesis a correlative formation of new functions and functional systems, including those that are not fully determined by the "phylogenetic laws." This idea, originating from the works by Severtsov and Schmalhausen, could account for evolutionary leap in the development of the brain in great apes, which led to rapid qualitative and quantitative changes resulted in the emergence of the perfect brain of modern men able to master and perform a variety of functions, including those that were not caused by long processes of phylogenetic development. These functions, in particular, include such aspects of abstract logical thinking as the ability to perform complex mathematical operations, create technical devices, and operate them at a distance. These functions also include those, which allow modern people to learn the principles of their own brain's organization. Based on our experimental data [50], we have suggested that speech could have evolved through the correlative phylogenetic development of stereognosis and language systems.

Thus, our recent data indicate that the implementation of complex verbal and mental activities associated with speech perception and production in children at different ages and in adults requires joint and extremely coordinated activities of both hemispheres along with an obligatory and differentiated involvement of the classical speech centers in the left hemisphere. These findings support the view on the role of interactions between both hemispheres as decisive for speech processing, as well as on the involvement of both hemispheres in the organization of different language levels. A quantitative analysis of the results obtained on children vs. adults allows assessing the degree of maturity of the organization of central processes underlying verbal-mnestic activity in preschool children. The new information about the formation of the lateralization of verbal processing in children's ontogenesis was obtained. The evidence for the essential role of joint activity of both brain hemispheres in providing speech was supported by the analysis of violations of the systemic organization of brain activity in children with impaired speech. The results of our research allow expanding modern concepts about the brain mechanisms of verbal activity in children in the norm and pathology. They can also assist in developing new and effective approaches to early correction of speech disorders.

Acknowledgements

This work was supported by state budget funding according to the assignment by the Russian Federal Agency for Scientific Organizations (FASO Russia) for Section 1 in 2018–2020 ("Mechanisms of formation of physiological functions in phylo- and ontogenesis as affected by endogenous and exogenous factors - № of state registration AAAA-A18-118012290373-7") and the means of the state budget for the state assignment in 2013–2017 for Section 3 "Comparative physiological and biochemical study of the processes of adaptation of humans and animals to changing conditions of natural, technological and social environment" (№ of state registration 01201351570).

Author details

Diana M. Guillemard (Tsaparina)*[1], Michail N. Tsitseroshin[1], Alexandr N. Shepovalnikov[1], Elizaveta I. Galperina[1], Ekaterina A. Panasevich[1], Ekaterina E. Kats[2], Larisa G. Zaytseva[1] and Olga V. Kruchinina[1]

*Address all correspondence to: diana-tsap@yandex.com

1 Sechenov Institute of Evolutionary Physiology and Biochemistry, Russian Academy of Sciences, St. Petersburg, Russia

2 Raul Wallenberg International University for Family and Child, Institute of Special Pedagogics and Psychology, Saint-Petersburg, Russia

References

[1] Shepovalnikov AN, Tsitseroshin MN. Formation of interzonal interaction of cortical fields during verbal-mnestic activity (in Russian). Zhurnal Evoliutsionnoĭ Biokhimii i Fiziologii. 2004;**40**:411-422

[2] Tsaparina DM, Tsitseroshin MN, Shepovalnikov AN. Reorganization of interhemispheric interactions during verbal-mental activity aimed at synthesis of words and sentences. Human Physiology. 2007;**33**:10-19

[3] Tsaparina DM, Tsitseroshin MN, Shepovalnikov AN. Age-related characteristics of the formation of neurophysiological mechanisms of the phonemic, grammatical, and semantic linguistic levels. Human Physiology. 2008;**34**:546-557

[4] Tsitseroshin MN, Tsaparina DM, Zaitseva LG. The role of interhemisphere interactions in children aged 5-6 years and adults in supporting verbal-mnestic activity associated with the formation and analysis of verbal utterances. Neuroscience and Behavioral Physiology. 2013;**43**:799-811

[5] Tsitseroshin MN, Pogosyan AA, Galperina EI, Shepovalnikov AN. Systemic interaction of the cortical areas during performance of verbal-mnestic activity. Human Physiology. 2000;**26**:665-674

[6] Povarova IA, Shepovalnikov AN, Tsitseroshin MN, Kats EE. Features of violations of systemic interaction between different areas of the cortex at stuttering. In: Povarova IA, editor. Stuttering: Diagnosis and Correction of Speech Tempo-Rhythmic Disorders (in Russian). Monograf. SPb: Speech. 2005. pp. 153-161

[7] Tsitseroshin MN, Shepovalnikov AN. Formation of the Integrative Function of the Brain (Book in Russian). SPb: Nauka; 2009. 249 p

[8] Sporns O, Tononi G, Kotter R. The human connectome: A structural description of the human brain. PLoS Computational Biology. 2005;**1**(4):e42

[9] Sporns O. Cerebral cartography and connectomics. Philosophical Transactions of the Royal Society of London. Series B, Biological Sciences. 2015;**370**:14

[10] Hagmann P. From diffusion MRI to brain connectomics [PhD thesis]. Lausanne: STI; 2005

[11] Alivisatos AP, Chun M, Church GM, et al. The brain activity map project and the challenge of functional Connectomics. Neuron. 2012;**74**:970

[12] Yuste R, Church GM. The new century of the brain. Scientific American. 2014;**310**:38

[13] Bola M, Sabel BA. Dynamic reorganization of brain functional networks during cognition. NeuroImage. 2015;**28**:1

[14] Craddock RC, Tungaraza RL, Milham MP. Connectomics and new approaches for analyzing human brain functional connectivity. Gigascience. 2015;**4**(13). DOI: 10.1186/s13742-015-0045-x

[15] Newton AT, Morgan VL, Rogers BP, Gore JC. Modulation of steady state functional connectivity in the default mode and working memory networks by cognitive load. Human Brain Mapping. 2011;**32**(10):1649-1659

[16] Kats EA, Galperina EK, Tsitseroshin MN. Disturbances of spatial organization of intercentral interaction in children with alalia and dysarthria (in Russian). Sensornye Sistemy. 2004;**18**:138-147

[17] Tsitseroshin MN, Ivonin AA, Pogosyan AA, et al. The role of the genotype in the development of the neurophysiological mechanisms involved in the spatial integration of the neocortex bioelectrical activity. Human Physiology. 2003;**29**:393-407

[18] Klimash AV, Tsitseroshin MN, Shepovalnikov AN, Zajceva LG, Kondakov EN, Borovikova VN. Disorders of the spatiotemporal organization of the Brain's bioelectrical activity in patients with different depressions of consciousness after severe head injury. Human Physiology. 2010;**36**:535-549

[19] Ivonin AA, Kutsenko DO, Tsitseroshin MN, Schepina AM, Shuvaev VT. Features of the spatial organization of biopotentials of the cerebral cortex in different clinical forms of neurotic depression (in Russian). Rossiĭskii Fiziologicheskiĭ Zhurnal Imeni I.M. Sechenova. 2010;**96**:259-269

[20] Hickok G, Poeppel D. Towards the functional neuroanatomy of speech perception. Trends in Cognitive Sciences. 2000;**4**:131-138

[21] Weiss S, Mueller HM, Rappelsberger P. Theta synchronization predicts efficient memory encoding of concrete and abstract nouns. Neuroreport. 2000;**11**:2357-2361

[22] Weiss S, Mueller HM. The contribution of EEG coherence to the investigation of language. Brain Language. 2003;**85**:325-343

[23] Chernigovskaya TV. The mirror brain, concepts, and language: The price of anthropogenesis (in Russian). Rossiĭskii fiziologicheskiĭ zhurnal imeni I.M. Sechenova. 2006; **92**(1):84-99

[24] Ross ED. Cerebral localization of functions and the neurology of language: Fact versus fiction or is it something else? The Neuroscientist: a Review Journal Bringing Neurobiology, Neurology and Psychiatry. 2010;**16**(3):222-243

[25] Dehaene-Lambertz G, Dehaene S, Hertz-Pannier L. Functional neuroimaging of speech perception in infants. Science. 2002;**298**:2013-2015

[26] Homae F, Watanabe H, Nakano T, Asakawa K, Taga G. The right hemisphere of sleeping infant perceives sentential prosody. Neuroscience Research. 2006;**54**:276-280

[27] Salvador R, Suckling J, Schwarzbauer C, Bullmore E. Undirected graphs of frequency-dependent functional connectivity in whole brain networks. Philosophical Transactions of the Royal Society B: Biological Sciences. 2005;**29**:937-946

[28] Buckner RL, Vincent JL. Unrest at rest: Default activity and spontaneous network correlations. NeuroImage. 2007;**37**:1091-1096

[29] Fox MD, Raichle ME. Spontaneous fluctuations in brain activity observed with functional magnetic resonance imaging. Nature Reviews. Neuroscience. 2007;**8**:700-711

[30] Ivanitskii GA, Nikolaev AR, Ivanitskii AM. Interaction of the frontal and left parietal-temporal cortex in verbal thought (in Russian). Fiziologiia Cheloveka. 2002;**28**:5-11

[31] Dan'ko SG, Kachalova LM, Solov'eva ML. Electroencephalographic characteristics of cognitively specific attention readiness in verbal learning. Communication 3. Topical characteristics of EEG spatial synchronization (in Russian). Fiziologiia Cheloveka. 2010;**36**:14-23

[32] Sakai KL. Language acquisition and brain development. Science. 2005;**310**:815-819

[33] Homae F, Watanabe H, Otobe T, Nakano T, Go T, Konishi Y, Taga G. Development of global cortical networks in early infancy. The Journal of Neuroscience. 2010;**30**:4877-4882

[34] Ries SK, Dronkers NF, Knight RT. Choosing words: Left hemisphere, right hemisphere, or both? Perspective on the lateralization of word retrieval. Annals of the New York Academy of Sciences. 2016;**1369**(1):111-131

[35] Simernitskaya EG. Human Brain and Mental Processes in Ontogeny (in Russian). Moscow: Mosk. Gos. Univ; 1985. 188 p

[36] Perani D, Saccuman MC, Scifo P, Anwander A, Spada D, Baldoli C, Poloniato A, Lohmann G, Friederici AD. Neural language networks at birth. Proceedings of the National Academy of Sciences of the United States of America. 2011;**108**(38):16056-16061

[37] Binder JR, Frost JA, Hammeke TA, et al. Human brain language areas identified by functional magnetic resonance imaging. Neuroscience. 1997;**17**(1):353-362

[38] Price CJ. The anatomy of language: A review of 100 fMRI studies published in 2009. Annals of the New York Academy of Sciences. 2010;**1191**:62-88

[39] Vigneau M, Beaucousin V, Herve PY, et al. Meta-analyzing left hemisphere language areas: Phonology, semantics, and sentence processing. NeuroImage. 2006;**30**:1414-1432

[40] Hagmann P, Cammoun L, Martuzzi R, et al. Hand preference and sex shape the architecture of language neurotransmitters. Human Brain Mapping. 2006;**27**:828-835

[41] Gevins AS, Zeitlin GM, Yingling CD, et al. EEG patterns during 'cognitive' tasks. I. Methodology and analysis of complex behaviors. Electroencephalography and Clinical Neurophysiology. 1979;**47**:693-703

[42] Pockberger H, Petsche H, Rappelsberger P, et al. On-going EEG in depression: A topographical spectral analytical pilot study. EEG. Clinical Neurophysiology. 1985;**61**:349-358

[43] Shepovalnikov AN, Tsitseroshin MN, Levinchenko NV. The developmental minimization of brain areas involved in the systems support for mental functions: Arguments for and against (in Russian). Fiziologiia Cheloveka. 1991;**17**(5):28-49

[44] Tsaparina DM, Shepovalnikov AN. The role of interhemisphere interactions in the process of recognizing errors in auditorily presented verbal material (in Russian). Sensornye Sistemy. 2004;**18**(2):160-169

[45] Yourganov G, Fridriksson J, Rorden C, Gleichgerrcht E, Bonilha L. Multivariate connectome-based symptom mapping in post-stroke patients: Networks supporting language and speech. The Journal of Neuroscience. 2016;**36**(25):6668-6679

[46] Blank I, Balewski Z, Mahowald K, Fedorenko E. Syntactic processing is distributed across the language system. NeuroImage. 2016;**127**:307-323

[47] Bastiaansen MC, van Berkum JJ, Hagoort P. Event-related theta power increases in the human EEG during online sentence processing. Neuroscience Letters. 2002;**323**:13-16

[48] Kruchinina OV, Galperina EI, Kats EE, Shepoval'nikov AN. Factors affecting the variability of the central mechanisms for maintaining bilingualism. Human Physiology. 2012;**38**:571-585

[49] Kruchinina OV, Galperina EI, Shepoval'nikov AN. Characteristics of the spatial organization of oscillations of brain bioelectric potentials in adolescents. Human Physiology. 2014;**40**:483-489

[50] Tsitseroshin M, Galperina EI. Correlative formation of functions as one of mechanisms of functional evolution (by example of development in child's ontogenesis of central maintenance of stereognosis and speech function). Journal of Evolutionary Biochemistry and Physiology. 2012;**48**:355-373

Chapter 8

Ontogenetic Development of Neural and Muscular Rhythmic Activity and Its Regulation in Mammals during Perinatal Period

Sergey Vladimirovich Kuznetsov,
Natalia D. Vdovichenko, Ludmila E. Dmitrieva,
Natalia N. Kuznetsova, Vladimir A. Sizonov,
Maksim A. Terpilowski and Olga P. Timofeeva

Additional information is available at the end of the chapter

http://dx.doi.org/10.5772/intechopen.73656

Abstract

This review covers our recent advantages in studying the ontogenetic aspects of physiological mechanisms underlying regulation of rhythmic behavior. We have revealed that excitation patterns that emerged at early stages of phylogenetic development of life forms contribute greatly to the rhythmic activity of living vertebrates and invertebrates. These patterns govern spontaneous excitation, which is easily observed during the early stage of ontogenesis. The intensity and patterns of rhythmic activity are determined by nature and kinetics of certain metabolic reactions. During perinatal and sometimes postnatal periods (as in prematurely born animals), endogenic rhythmicity of developing physiological structures is strongly pronounced due to relatively stable living conditions. This rhythmic behavior is coordinated within an entire organism. Its integration in multiple systems is driven by amplitude and frequency modulation yielding rhythms of various frequency ranges. Indeed, it is the complex and conjoint functioning of physiological systems that maintains homeostasis in developing organisms. We present the results of our authentic research concerning the evolution and ontogeny of regulatory mechanisms of motor, cardiovascular, and respiratory systems. The aspects of intact and disrupted development are considered, involving the changes in dopaminergic, norepinephrinergic, and cholinergic system activation.

Keywords: evolution physiology, perinatal development, biorhythms, metabolism, neuropharmacology, heart rhythm, motor activity, respiration

1. Introduction

Endogenic rhythmic activity prevails during prenatal and—in immature animals—postnatal periods due to relatively stable living conditions. This rhythmic activity is strongly coordinated on an organism level. Its integration is done through amplitude and frequency modulation which yields rhythms of several frequency ranges. Only conjoint activity of all systems provides the main life-supporting factor—homeostasis of a developing organism.

The concept of archetypal "primary rhythms" was formulated by Voĭno-Yasenetskii A.V. as a result of long-term physiological research. Excitation rhythms detected in a complex organism or a cell located in a non-oscillating environment are considered primary rhythms. "Development of periodic nature of rhythmic excitation of cells is a result of subordinacy of biochemical processes kinetics to periodic impact of external environment during the early stages of life development" [1]. Primary rhythms are not exclusively attributable to primitive life forms or early developmental stages of living organisms. Complexes of periodic activity also persist in higher animals during the ontogenetic development and maturation of reticular systems, involved in synchronization of other brain structures. Considering the principles of CNS evolution during ontogenesis of vertebrates, Voĭno-Yasenetskii emphasized that it is the rhythm that serves a basis for CNS motor function development, and brain development in all warm-blooded animals follows the same principle of evolution [1].

We hold onto a classification of biorhythms, proposed by J. Salanki, who divided biological rhythms into three categories: micro, meso, and macrorhythms [2]. Rhythms which fall into a mesogroup and have a period between 1 s and several minutes are of the greatest interest for us. Excitable structures of organisms including Protozoa and higher animals produce primary rhythms which belong to this group. It should be noted that endogenic rhythms of various frequency ranges often coexist and are to be considered in a robust relationship.

The problem of endogenic rhythms origin is very complicated and is not resolved yet. We could not provide even a superficial review of available hypotheses and scientific data within the scope of this chapter. Therefore, we briefly discuss several publications concerning periodic nature of biochemical and biophysical processes. It was shown that glycolysis oscillation period falls into the 1–3-min interval; succinate oxidation has a period of 15 s, and ATPase rhythm has a period of 3–5 s [3–6]. Analysis of background oscillations of neuron cytoplasmic microstructures has revealed rhythms with periods of 1–2 s (movement and aggregation of mitochondria, chromatin pulsation) and 15–30 s (change of distance between axon hill and nuclear) [7]. The research carried out on isolated crawfish mechanoreceptors has helped register oxygen consumption oscillations with a period of 11–18 s, where amplitude increases dramatically on cell functional state change. Spectral analysis helped establish several groups of discrete periods of oscillations of mitochondria aggregations in various functional states of a neuron: 3–5, 11–18, 28–40 s, and 1–3 min. Time parameters of cellular energetics are probably associated with metabolic energy expenditure. "Superslow rhythmic oscillations of action potential," according to Aladzhalova N.A. terminology [8], or "slow electric processes," as defined by Ilyukhina V.A. [9], were found and extensively studied in cortex and deep structures of the brain in warm-blooded animals. These groups of potentials embraced rhythmic oscillations with periods between 2 s and 6 h. It was detected that sleep spindles in newborn rabbit cortex have a period of 10–15 s and coincide with motor activity [10]. Rhythms with period ranges considered in this chapter were classified by researchers into near-second (2–12 s), decasecond (15–20 s), near-minute rhythms (1 min) [8], or ζ-waves (2–10 s), τ-waves (12–60 s), ε-waves (1–5 min). Each range includes several subranges with varying physiological features [9].

Currently, there are no generally accepted time intervals for rhythm periods of given frequency ranges. Scientific data obtained in our research let us classify them into the following groups: near-second rhythms with a period of 0.5–3.5 s, decasecond rhythms (4.5–45 s), near-minute rhythms (50–120 s), and multiminute rhythms (125–600 s). This classification is not robust, because rhythm periods may vary greatly due to certain functional changes (periods sometimes shift quite considerably to an adjacent frequency range) [11]. We avoid contiguously defining the rhythm periods, because rhythm lability is an indispensable feature of an excitable structure.

2. Development of rhythmic functions during prenatal ontogenesis in rats

It is well established that endogenic rhythms of near-second and near-minute periods—the basis of spontaneous excitation—have relatively close origins, independent of the excitable structure or a functional system they belong to. Our data, backed by other researches, have shown that rhythms of the same frequency range detected in various functional systems of an organism are tightly adjacent by their ontogenetic dynamics and share regulatory, mediatory, and metabolic status on a certain developmental stage. Ontogenetic trends of rhythm timing parameters are most marked in near-second rhythms, which are detected in the gastrointestinal tract, motor, and respiratory activity. Rhythm period decreases soon after birth and up to postnatal day 16 (P16). For instance, a respiration rhythm decreases from 1.3 to 0.4 s and a locomotor rhythm—from 1.4 to 0.7 s. Later, gut rhythmic segmentation period declines from 3.5 to 2.2 s, probably due to changes in feeding behavior. Decasecond and near-minute rhythms do not undergo such dramatic changes, and their timing parameters do not alter over time. It is probably related to two aspects: (1) high rhythm variability is due to a wide frequency range and (2) such rhythms often have a modulating activity, that is, generated by early maturating structures [11, 12].

Spontaneous periodic motor activity is a dominant type of motor activity within prenatal and early postnatal periods of ontogenesis. Its periods are conventionally classified into near-second, decasecond, and near-minute periods (**Figure 1**). It was shown that heart, respiratory, and motor activities in rat fetuses with normal blood circulation change on embryonic days 15–20 (E15–20). Motor activity of rat fetuses has a complicated periodic pattern, resulting from three main types of locomotion. The first type corresponds to a generalized activity—body flexions and limb movement. Generalized activity peaks on E18 and completely

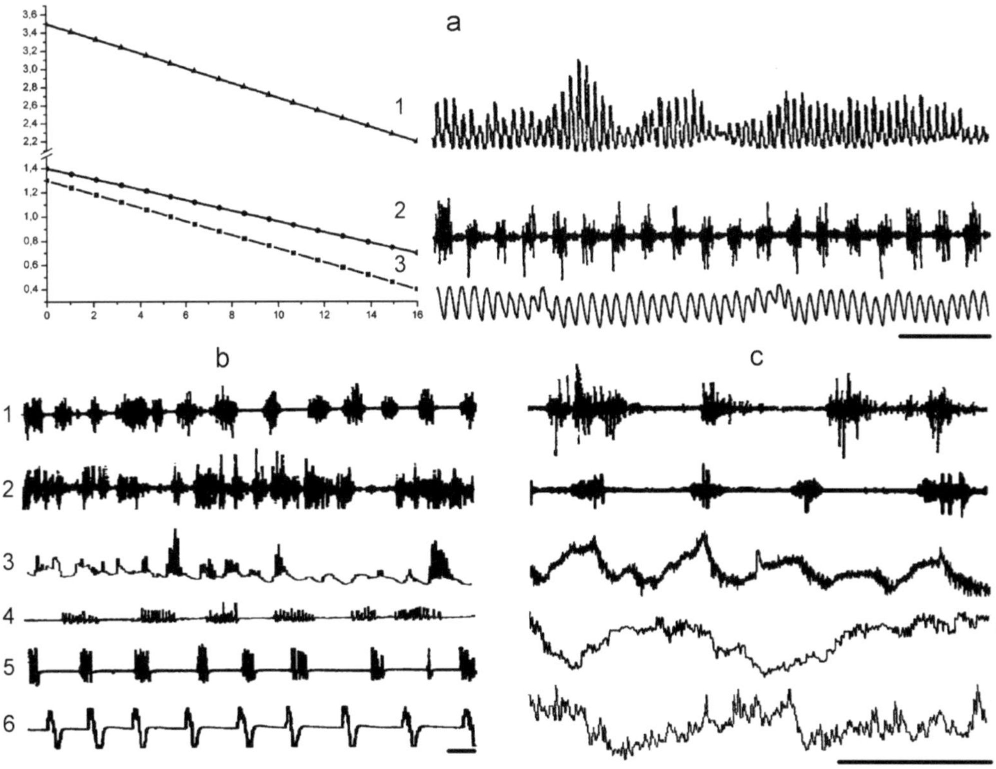

Figure 1. Examples of rhythms of various frequency ranges. Ontogenetic dynamics (a, on the left) of period of near-second rhythm and its reproduction (a, on the right) in pattern activity of gastrointestinal tract (1), SPMA (2), and respiration (3). Abscissa: Age of rat pups (days), ordinate: Period of rhythm (s). Time bar for electrograms: 10 s. (b) the appearance of decasecond periodicity present in SPMA of the 5-day-old rat pup (1), SPMA of the 2-day-old rat pup after administration of isoamine at a dose of 3 mg/kg (2), spontaneous contractions of isolated duodenum after addition to incubation solution of lysenyl at a concentration of 3×10^{-8} M (3), mechanogram of the 5-day-old rat pups after administration of hydrogen peroxide at a dose of 125 mg/kg (4), respiratory rhythm of 0-day rat pup after administration of lobeline at a dose of 50 mg/kg (5), respiration of adult human (6). (c) The appearance of the near-minute activity rhythm in SPMA of the 5- (1) and 10-day-old (2) rat pup, cardiointervalogram of the 5-day-old rat pup (3), spontaneous contractions of isolated ileum of the 30-day-old rat pup (4), and spontaneous contractions of isolated duodenum of adult rat (5). Time bar (s): (a), (b)—10, (c)—30.

disappears on E20. Burst duration increases initially from 17.1 ± 1.8 s on E16 to 36.7 ± 12.0 s on E18 and then decreases to 9.7 ± 1.0 s on E20. The second type of locomotion is related to local body movements which have a duration of about 5 s. They may produce series of 3–7 movements occurring in a decasecond range. The third type arises as short—about 300 ms—irregular extensor jerks. Local movements and jerks are typical of all ages, but they appear to be the main type of activity on E20. The development of respiratory activity sees a gradual transition from separate breaths on E16 to a series of breathing movements on E17–18 and then to episodes of periodic respiratory activity.

The analysis of interactions between respiratory and locomotor systems has shown that every breathing movement in 16-day fetuses is accompanied by an extensor jerk. Respiratory and locomotor activities are mostly uncoupled by E19. Heart rate increases from 175.9 ± 6.1 on

E15 to 271.8 ± 5.9 bpm on E20. Newborn rats have a volatile respiratory activity with considerable variations of rhythm and amplitude. Sometimes apnea periods are detected in rat puppies. Respiratory rhythm is strongly associated with other aspects of respiratory activity. Destabilization of respiratory rhythm accompanies not only complexes of motor activity but also single jerks. Respiratory rate variations which are related to SMA and observed at various ages have different trends. Rat puppies on P0-1 exhibit a tendency toward decreasing respiratory rate during rest (between complexes of locomotor activity), wherein a burst of locomotor activity is accompanied by acceleration of respiratory rate. SMA intensity declines by 2-week age; respiratory rate increases at rest and decreases during the burst of motor activity [13]. SMA which is detected in rat puppies during early postnatal ontogenesis consists of bursts of motor activity and rest periods. Activity-rest complex usually has a decasecond (40–45 s) or a near-minute rhythm. Decasecond rhythms can also be components of near-minute rhythms. Near-second rhythm as a basis of locomotor activity in mature animals is observed within bursts of motor activity with decasecond and near-minute rhythms. Near-minute rhythm of SMA is the most pronounced during early postnatal ontogenesis, while decasecond rhythm begins to prevail in the course of maturation. On acquiring a sense of sight and at the end of breastfeeding phase, periodic complexes of motor activity are replaced by near-second locomotor rhythm [14, 15].

Rhythmic activity of a certain frequency range can be simultaneously triggered by multiple physiological systems. Comparable rhythm parameters detected in many structures imply that there are common mechanisms of oscillation. The main candidate is a shared metabolic substrate. It is well known that evolution favors structures with slow oxidative phosphorylation, because it makes them less vulnerable to environmental changes. Phylogenetically oldest structures rely mostly on pentose phosphate cycle. They are able to retain automatic rhythmic functioning even in unfavorable conditions [16].

Our research has revealed that inhibitors of pentose phosphate pathway enhance the activity of physiological systems being studied. For instance, administration of hydroquinone to rat puppies completely eliminates rest from a typical activity-rest cycle of SMA, increases duration and amplitude of locomotor activity, and minimizes modulating rhythms [17]. Glucose is catabolized not only in pentose phosphate pathway but also mostly in glycolysis and citric acid cycle. Therefore, we carried out a research involving inhibitors of citric acid cycle. Newborn rats who were given sodium fluoroacetate intravenously died within a week: cachexia and developmental growth failure were observed. Heart and respiratory functions were not impaired in puppies unlike mature rats. This fact indicates that excitable structures are less dependent on aerobic metabolism at early ontogenesis. It is also supported by the absence of SMA arrest in test animals after administration of inhibitors. Citric acid cycle inhibition affects amplitude but not the pattern of SMA. The analysis of experimental and theoretical data suggests that cholinergic system plays a key role in physiological regulation when aerobic pathways are inhibited. Thus, a set of features specific to phylo- and ontogenetically early stages is manifested. Given that endogenic rhythmic activity is strongly tied with pentose phosphate cycle, we propose a hypothesis that citric acid cycle inhibition leads to redirecting carbohydrates to pentose cycle in excitable structures involved in excitation propagation [18].

Thus, we have established that glycolysis plays an important role in motor activity of newborn rats. Further research has revealed that bursts of SMA consisting of decasecond and near-minute rhythms are accompanied by a drop in serum glucose level. It should also be noted that activity pattern is affected by a feed state: emptying of stomach is followed by an arousal of decasecond rhythms and rapid jerks. Glucose level decreases by approximately 25% during rest and by 25–40% during activity. We have not detected any age- and pattern-related diversity. There is a seasonal variability in serum glucose level in newborn rats. However, there is a strong correlation between glucose level and motor activity pattern during all seasons except spring. The differences are smoothened in spring [19, 20].

Analysis of physiological parameters—heart and respiratory rate—in intact newborn rats has also shown a high correlation with season and feeding state. We also studied SMA parameters and blood glucose level in fed and fasting states (after a day-long starvation). Overall, blood glucose level in starved rats was 1.5–2.5 times lower than that in fed animals. Administration of glucose to fasting and—to a less extent—fed animals of all ages led to potentiation of near-minute rhythm of activity. Postnatal day 0 is an exception, as administration of glucose to fed animals intensifies decasecond rhythm and decreases overall motor activity. Blood glucose level and ontogenetic dynamics of SMA parameters in fasting rat puppies show a significant correlation that is not observed in fed animals: blood glucose level is higher during activity than at rest. Revealed ontogenetic variability of response toward glucose tolerance test is probably induced by morphofunctional maturation of locomotor system and carbohydrate metabolism juvenility. Our results suggest that intensity, duration, and pattern of SMA in newborn rats strongly depend on feeding state and can vastly change due to induced hypo- and hyperglycemia [21].

3. Development of cross-system interactions in perinatal ontogenesis of rats

The exact mechanisms of development and interactions of modulating rhythms remain unclear, but these problems are essential for ontogenetic physiology. It is well known that interactions between heart-vascular and respiratory systems are crucial for normal functioning of an organism. During prenatal ontogenesis, SMA plays a substantial role in morphogenesis and homeostasis. Functional activity of these systems has a rhythmic nature, and its coordination is crucial for homeostasis and adaptation to external and internal factors. In this section, we review some aspects of systemic interactions and regulation during late prenatal and early postnatal periods of ontogenesis.

We have obtained evidence that cardiorespiratory and viscerosomatic interactions emerge during prenatal period (E17–20) in rat fetuses. Tachycardia accompanied by R-R-intervals fluctuations develops during episodes of rhythmic respiratory activity. We have observed heart rate variability (HRV) in decasecond and near-minute ranges. Analysis of functional activity of cardiovascular and somatomotor systems has revealed that heart rate variability and burst of locomotor activity are independent. HRV becomes more synchronized with motor activity during maturation. Short-term heart decelerations associated with motor activity are specific to E17–18. Their duration and amplitude are weakly dependent on locomotion intensity. Decelerations are superseded by acceleration reactions that are typical of a mature organism on E19–20. It is related to development of coordinating function of nervous system [22].

We have analyzed slow-wave oscillations of heart rhythm and motor activity in fetuses for any correlations with mother's heart rate in three frequency ranges: decasecond, D1–0.02–0.2 Hz; near-minute, D2–0.0083–0.02 Hz, and multiminute, D3–0.0017–0.0083 Hz. Pearson coefficient of correlation, its sign, and time shift have been estimated as correlation parameters. No correlation has been found in D1 range without regard to age. Correlations are higher in D2 and D3 ranges. Maximum correlation has been detected on E18–19, when motor activity peaks. Respiratory and cardiovascular systems of mother affect the variability of rhythmic processes in fetuses to a certain extent. Heart rates of mother and fetus strongly correlate with E17 and E20, when motor activity is at its breakpoint–rising and falling, respectively. In most cases, heart rate variability of mother outruns variability of fetus in D2 and D3 ranges. Specific mechanisms of heart rate synchronization between mother and fetus are not known yet. There are two hypotheses: (1) similar oscillators with close parameters in mother and fetus and (2) mother's rhythm affecting fetus directly [23].

An example of synchronizing slow-wave constituents of rhythmic activity of mentioned functional systems is given in **Figure 2**.

Interactions between physiological systems through slow-wave rhythms manifest themselves more completely during postnatal ontogenesis. Respiratory system in fetus changes dramatically after birth, interfering with other excitable structures. During first hours after birth, interactions between somatomotor and cardiovascular systems strengthen. Accelerations are typical of newborn rats, being greatly affected by intensity of motor burst. Interactions mediated by D2 and D3 rhythms are most pronounced in intact rat puppies: extent of cooperation goes down in the following pairs—somatomotor and cardiovascular systems, somatomotor and respiratory systems, respiratory and cardiovascular systems. Correlation decreases from D3 to D1 range, D3 having a highest correlation. D1 rhythms do not participate in intersystemic interactions. Changes in SMA intensity and pattern have a tendency to outrun modulating oscillations in D2 and D3 ranges of heart and respiratory rhythms [13, 24]. Consequently, systemic interactions in newborn rats are accomplished mostly by slow-wave oscillations of D2 and D3 ranges. Rhythms of decasecond range, D1, do not play an important role in integrative processes.

These mesorhythms were detected in fast temperature fluctuations in newborn rats up to 1 week of postnatal development. It was observed that body temperature fluctuated with an amplitude of 0.04–0.09°C, which corresponded to near-second and near-minute rhythm ranges. These temperature fluctuations coincide with rhythmic excitation in other functional systems. There is a clear correlation between bursts of SMA and temperature oscillations. Simultaneous recording of rectal temperature, respiratory rate, SMA, and electric activity of stomach established that temperature oscillations were more frequent than motor bursts or stomach activity. High-amplitude spiking activity of stomach muscles surpasses complexes

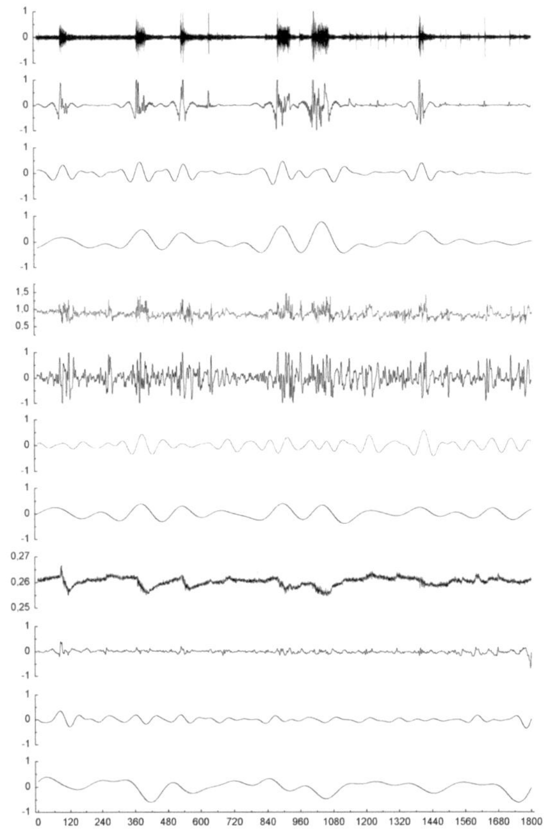

Figure 2. An example of production of initial signals. In (a)—EMG filtration (1), periodograms of the respiration rhythm (2), periodograms of the cardiac rhythm (3) to the slow-wave components (diapasons D1, D2, and D3). Abscissa: time (s), ordinate: the amplitudes of EMG signals, D1, D2, and D3 (stand. units); the periodograms of respiration and cardiac rhythms (s).

of motor activity and fits temperature peaks. Temperature fluctuations are unlikely to play a key role in thermal regulation due to low amplitude and high frequency. Most probably, these oscillations are mediated by systemic vasomotor reactions. Such temperature variations may indicate metabolic dynamics and may serve as a trigger to rhythmic excitation. They may also be involved in synchronization of rhythmic activity in various structures [25].

4. Monoaminergic regulation of rhythmic activity during perinatal ontogenesis in rats

Maintaining relevant rhythmic activity of physiological systems is essential for normal functioning and homeostasis. It is well known that rhythmic activity of visceral and somatic organs or systems of organs includes not only inner rhythms but also the activity of interacting structures (outer rhythms). Rhythmic activity integration is accomplished with its amplitude and frequency modulation by rhythms of higher orders. Integration pattern is related to development of regulatory systems. It was shown that alterations in thresholds of cholinergic and adrenergic systems could result in severe functional disorders including cardiovascular and respiratory abnormalities [15].

Catecholaminergic system is notable for its early ontogenetic development. However, research carried out on fetuses is limited. This fact complicates analysis of underlying processes leading to dysfunction development in newborn animals. We have conducted a pioneering physiological research in rat fetuses (E17–20) and in newborn rats (P0–1) in vivo. Results included cardiovascular, respiratory, and somatomotor activity parameters in dynamics. The experiments were conducted in intact and test animals that were exposed to physiological and pharmacological activation of dopaminergic and noradrenergic systems [26–29].

L-DOPA was administered to animals to increase catecholamine levels in doses of 25, 50, and 100 mg/kg. The activation of catecholaminergic structures was induced by administration of isoamine (an indirect adrenergic agonist) in doses of 3 and 10 mg/kg. Moreover, we studied action of L-DOPA and isoamine after blockade of D1-receptors (by an antagonistic drug SCH-23390, 0.1 mg/kg) and D2-receptors (sulpiride, 50 mg/kg).

It was discovered that injection of L-DOPA resulted in continuous generalized motor activity in dose-independent (E17–18) and dose-dependent manner (E19–20). Respiratory rate in fetuses increases after L-DOPA injection. The number of fetal respiratory movements (gaspings) increases by 3–7 times. Heart rate is not affected considerably by L-DOPA. Our experiments that involved clonidine, an α-adrenergic agonist (1 mg/kg), administration to fetuses show that noradrenergic system of fetuses is capable of stimulating respiratory activity.

Novel head movements having a near-second rhythm appear between E18 and E19 during ontogenesis in 92% of fetuses. Mature activity patterns emerge in response to L-DOPA after E19: stereotypical head movements (circular movements, lateral and dorsoventral flexions) and alternating forelimb movements. Effects of L-DOPA derivatives—noradrenaline and dopamine—were brought into question. Based on theoretical and experimental data obtained by us and other authors, we expected L-DOPA to activate noradrenergic system. Stereotypical behavior in mature animals is currently associated with dopaminergic system by many authors. It seems well reasoned to suggest that dopaminergic system is also responsible for stereotypy in fetuses. We studied the effects of dopamine antagonists in fetuses on E19–20 and revealed that D1- and D2-receptor blockade did not affect motor activity of fetuses. L-DOPA induced the same effects when given with inhibitors. It implies that stereotypy in fetuses is not linked to dopamine system. Dopamine antagonists do not alter gasping and do not block L-DOPA effects on respiratory activity. DOPA can act as a CNS neurotransmitter binding to D2 and β2-adrenergic receptors [30]. We studied propranolol effects and came to conclusion that this phenomenon might be linked to noradrenergic system.

Endogenic monoamines triggered by isoamine tend to stimulate motor activity in fetuses on embryonic days 17–18. Isoamine induces two types of reactions on E19–20: a short activation of motor activity followed by inhibition (in 60% fetuses) and overall inhibition (in 40% fetuses). It is likely that noradrenergic regulation is altered on E19–20 (before birth). Excessive concentration of catecholamines favors an increase of motor activity. Endogenic catecholamine release is followed by a short-term stimulation and then inhibition. Activation of catecholaminergic pathways leads to significant increase of motor activity during postnatal ontogenesis. Isoamine augments respiratory movements by two times on E17–18. We have observed an activation of respiratory activity in 60% fetuses and reduction in 40% fetuses on E19–20. Heart rate rises slightly on E17–18 and then decreases on E19–20. Heart rate and respiratory rate decrease slightly in newborn rats on administration of isoamine. D1- and D2-receptor blockade does not alter motor activity, respiratory, and heart rates of fetuses. Subsequent physiological or manual activation of catecholaminergic systems results in multiple reactions due to ontogenetic and individual features of animals.

Newborn rats are also subject to dose-dependent increase of SMA by L-DOPA. Stereotypical head movements (up-down, from side to side), body movements, and alternating limb movements are observed. Motor bursts occurring in a decasecond rhythm are significantly enhanced. Two-week rat puppies are also subject to increase of motor activity triggered by L-DOPA. However, this effect is less pronounced than in newborn rats. Locomotor rhythms are dominant in such motor activity.

Catecholaminergic system-induced respiration effects change remarkably on P0–1: excessive catecholamines release results in an increase of respiratory movements (gaspings) during prenatal period, and respiratory rhythm slows down on P0–1. Release of catecholamines potentiates SMA in newborn rats. D1-receptor blockade stimulates SMA. D2-receptor antagonists slightly augment SMA. L-DOPA further stimulates motor activity and changes its pattern. We detected oscillation complexes which had a period of about 3–6 min and tendency to expand and shift to a continuous mode. Motor complex patterns consist mostly of locomotor rhythms. Stereotypical movements are preserved. Isoamine injected after D2-receptor blockade had a weaker effect than that injected after L-DOPA administration.

Heart rhythm of intact newborn rats is a sinus rhythm with amplitude and frequency modulation. Single or multifocal extrasystoles are occasionally detected. Slow-wave modulation of heart rhythm in intact rat puppies consists of frequency oscillations falling into decasecond and near-minute ranges. These modulations are subdivided into asynchronous (mostly) and synchronous (with SMA complexes), the latter bearing a tachycardial pattern. Moreover, irregular bradycardial decelerations unrelated to motor activity are observed in heart rhythm. These fluctuations have a duration of about 10–30 s and an amplitude of 130–150% to a mean rhythm period of heart rate. D1-receptor blockade does not alter heart rate, while D2-blockade decreases heart rate. L-DOPA lowers respiratory rate when administered to newborn rats on P0–1. D1-receptor blockade does not change respiratory rate, but subsequent injection of L-DOPA leads to irregular respiration rhythm and low respiratory rate. D2-receptor blockade decreases respiratory rate, but this effect is opposed by L-DOPA. Stimulation of amplitude modulation is accompanied by an irregular respiratory rhythm.

We would like to draw attention to the data obtained during heart rate variability analysis. Spectrum power alterations in various frequency ranges were analyzed against physiological parameters registered on administration of various drugs. It was revealed that newborn rats had unidirectional changes in spectra related to activity of catecholaminergic systems, while rat fetuses (E17–18) displayed a large variability due to the influence of multiple regulatory systems. During prenatal period, L-DOPA enhances neural regulation of vagosympathetic balance in 50% rat fetuses, which is caused by alterations in humoral and metabolic factors. In the rest 50% fetuses spectrum power is shifted toward slow-wave activity. Moreover, in 75% newborn rats, L-DOPA enhances high-frequency constituents of spectra, that is, humoral and metabolic influence is minimized. The observed changes involved activation of both sympathetic and parasympathetic systems. But parasympathetic effects are more pronounced, which is reflected by LF/HF coefficient means. Isoamine boosts sympathetic effects by 70–75%. It was revealed that neural regulatory mechanisms began to play a more important role in the regulation of heart rhythm in newborn rats. Sympathetic effects were observed in 60% animals. However, vagosympathetic balance shifts toward parasympathetic effects due to their prevalence in overall spectra.

So, it is known that D1 and D2 receptors are subject to heterochrony (in terms of physiological development). Earlier it was proposed that D1 receptors are physiologically inhibited within 30 days of postnatal ontogenesis [31]. This fact is backed by another research [32, 33]. But herein we show that D1- and D2-receptor blockade affects respiratory activity greatly on E17. Moreover, there are differences in the response to dopamine receptors antagonists in fetuses on E17–18 and E19–20. Similar results were obtained earlier as a result of motor activity analysis in fetuses on E21 with D1- and D2-receptor antagonists [33]. Based on our results, we suggest that catecholaminergic systems undergo significant changes during prenatal ontogenesis (E18–19). It is supported by the fact that an increase of catecholamines on E19 is accompanied by appearance of stereotypical movements, which is not inhibited by dopamine receptors and β2-adrenoreceptor blockade.

Thus, we have shown that effects of catecholaminergic and dopaminergic systems on heart and respiratory regulation change dramatically after the birth. Catecholaminergic system changes are the most pronounced. Respiratory system response to physiological and excessive concentrations of endogenic monoamines alters within a few hours in newborn rats. The variability of observed response depends on several factors: age, feeding state, and season. The significance of two latter ones was demonstrated in our recent work involving complex biochemical and physiological methods [19]. The key role of functional changes occurring in respiratory system and catecholaminergic structures activation is supported by our research dedicated to systemic interactions and reviewed in Section 3.

Acknowledgements

This work was supported by the state budget funding according to the assignment by the Russian Federal Agency for Scientific Organizations (FASO Russia) "Formation mechanisms of physiological functions in phylo- and ontogenesis as affected by endogenous and exogenous factors."

Author details

Sergey Vladimirovich Kuznetsov*, Natalia D. Vdovichenko, Ludmila E. Dmitrieva, Natalia N. Kuznetsova, Vladimir A. Sizonov, Maksim A. Terpilowski and Olga P. Timofeeva

*Address all correspondence to: ksv@iephb.ru

The Laboratory of Ontogenetic Development of Animals Nervous Activity, I.M. Sechenov Institute of Evolutionary Physiology and Biochemistry RAS, St. Petersburg, Russia

References

[1] Voĭno-Iasenetskii AV. Primary Excitation Rhythms During Ontogenesis. Leningrad: Nauka; 1984. 148 p. (in Russian)

[2] Salanki J. Neural mechanisms in rhythm regulation of invertebrates. Neurobiology of Invertebrates. 1971;**1973**:17-31

[3] Kondrashova MN, Kaminskiy YG. Oscillatory Processes of Mitochondria Due to Potassium Acetate Transport. 2nd ed. Pushchino: Pushchinskii nauchnyi tsentr; 1971. pp. 72-75. (in Russian)

[4] Israel V, Lesbats B, Marsal J, Meunier FM. Oscillation of adenosine triphosphate and acetylcholine levels during stimulation of the electric organ of the torpedo fish. Comptes Rendus Hebdomadaires des Seances de l'Academie des Sciences Serie D. 1975;**280**(7):905-908

[5] Hess B. Oscillations in biochemical systems. Life Sci. Res. Rep. 1976;**1**:175-191

[6] Hess B. The glycolytic oscillator. The Journal of Experimental Biology. 1979;**81**(1):7-14

[7] Zaguskin SL, Nikitenko AA, Ovchinnikov YA, Prokhorov AM, Savranskii VV, Degtiareva VP, Platonov VI. On period range of oscillations in microstructures of a living cell. Doklady Akademii Nauk SSSR. 1984;**277**(6):1468-1471. (in Russian)

[8] Aladzhalova NA. Psychophysiological Aspects of Super Slow Rhythmic Activity of Brain. Moskva: Nauka; 1979. 214 p. (in Russian)

[9] Iliukhina VA. Neurophysiology of Human Functional States. Leningrad: Nauka; 1986. 289 p. (in Russian)

[10] Voĭno-Iasenetskii AV, Dmitrieva LE. Synchronization and desynchronization of cerebral rhythms of activity during ontogenesis. Zhurnal Evoliutsionnoĭ Biokhimii i Fiziologii. 1973;**9**(4):390-397. (in Russian)

[11] Kuznetsov SV. To the problem on nature and origin of ancient rhythms of excitation. Journal of Evolutionary Biochemistry and Physiology. 1999;**35**(5):456-467

[12] Bursian AV, Dmitrieva LE, Kulaev BS, Polyakova LA, Semenova YO, Sizonov VA, Timofeeva OP. Synchronization of endogenous somato- and visceromotor activities in rats in ontogenesis. Journal of Evolutionary Biochemistry and Physiology. 2001;**37**(3):258-265

[13] Kuznetsov SV, Sizonov VA, Dmitrieva LE. Change of character of intersystemic interactions in newborn rat pups under conditions of a decrease of central influences (urethane anesthesia). Journal of Evolutionary Biochemistry and Physiology. 2014;**50**(3):234-246

[14] Bursian AV, Timofeeva OP. The formation of activity-rest cycles in early ontogeny. Zhurnal Evoliutsionnoĭ Biokhimii i Fiziologii. 1990;**26**(1):85-91. (in Russian)

[15] Kuznetsov SV. About mechanisms of genesis, structure, and functional role of endogenous rhythms (to the 100-anniversary of Aleksei Valentinovich Voino-Yasenetskii). Journal of Evolutionary Biochemistry and Physiology. 2009;**45**(6):643-662

[16] Laborit H. Les régulations métaboliques. Paris: Masson & Cie; 1965. 498 p

[17] Kuznetsov SV. The dependence of spontaneous excitation processes on metabolic activity in the early postnatal period. Journal of Evolutionary Biochemistry and Physiology. 1996;**32**(4):338-345

[18] Kuznetsov SV. Study of ECG, parameters of external respiration and motor activity of rats of different ages during acute poisoning with fluoroacetamide. Journal of Evolutionary Biochemistry and Physiology. 2006;**42**(4):431-445

[19] Kuznetsov SV, Selina EN, Kuznetsova NN. The blood glucose content in newborn rats depending on level and pattern of spontaneous motor activity. Journal of Evolutionary Biochemistry and Physiology. 2011;**47**(4):374-380

[20] Kuznetsova NN, Selina EN. Fluctuation of blood glucose level in newborn rats depending on the season and spontaneous motor activity pattern. Journal of Evolutionary Biochemistry and Physiology. 2015;**51**(5):396-400

[21] Kuznetsov SV, Kuznetsova NN, Shpakov AO. Interconnection between parameters of motor activity and blood glucose concentration in newborn rats at starvation and under glucose load conditions. Journal of Evolutionary Biochemistry and Physiology. 2014;**50**(4):321-333

[22] Timofeeva OP, Vdovichenko ND, Bursian AV. Quantitative assessment of the relationship of slow-wave heart rate oscillations and motor activity in rat fetuses with maternal respiratory and cardiac activity. Journal of Evolutionary Biochemistry and Physiology. 2015;**51**(5):412-420

[23] Vdovichenko ND, Timofeeva OP. Quantitative estimation of connection of the heart rate rhythm with motor activity in rat fetuses. Journal of Evolutionary Biochemistry and Physiology. 2014;**50**(2):103-108

[24] Sizonov VA, Dmitrieva LE, Kuznetsov SV. Somatovisceral intersystemic interactions in newborn rat pups under decreased motor activity. Journal of Evolutionary Biochemistry and Physiology. 2015;**51**(4):332-341

[25] Kuznetsov SV. Rapid temperature fluctuations as a synchronization of processes of spontaneous excitation in rat pups. Journal of Evolutionary Biochemistry and Physiology. 1999;**35**(3):376-383

[26] Kuznetsov SV, Dmitrieva LE, Sizonov VA. Cardiac, respiratory, and motor activity in norm and after activation of catecholaminergic systems in newborn rat pups. Journal of Evolutionary Biochemistry and Physiology. 2012;**48**(4):430-444

[27] Timofeeva OP, Vdovichenko ND, Kuznetsov SV. Dynamics of the formation of rhythmic activity of the heart in fetuses and newborn rats. Bulletin of Experimental Biology and Medicine. 2012;**152**(4):397-401

[28] Timofeeva OP, Vdovichenko ND, Kuznetsov SV. Effect of change in activity level of catecholaminergic systems on motor, respiratory, and cardiac activities in fetal rats. Journal of Evolutionary Biochemistry and Physiology. 2012;**48**(3):310-321

[29] Timofeeva OP, Vdovichenko ND, Kuznetsov SV. Change in character of motor, respiratory, and cardiac activities in rat embryos at stimulation of release of endogenous catecholamines. Journal of Evolutionary Biochemistry and Physiology. 2012;**48**(3):346-350

[30] Misu Y, Kitahama K, Goshima Y. L-3,4-Dihydroxyphenylalanine as a neurotransmitter candidate in the central nervous system. Pharmacology & Therapeutics. 2003;**97**(2): 117-137

[31] Chen YI, Choi JK, Xu H, Ren J, Andersen SL, Jenkins BG. Pharmacologic neuroimaging of the ontogeny of dopamine receptor function. Developmental Neuroscience. 2010;**32**(2):125-138

[32] Moody CA, Spear LP. Ontogenetic differences in the psychopharmacological responses to separate and combined stimulation of D1 and D2 dopamine receptors during the neonatal to weanling age period. Psychopharmacology. 1992;**106**(2):161-168

[33] Moody CA, Robinson SR, Spear LP, Smotherman WP. Fetal behavior and the dopamine system: Activity effects of D1 and D2 receptor manipulations. Pharmacology, Biochemistry, and Behavior. 1993;**44**(4):843-850

Chapter 9

Development of the Biosphere in the Context of Some Fundamental Inventions of Biological Evolution

Vladimir F. Levchenko, Alexander B. Kazansky and Marat A. Sabirov

Additional information is available at the end of the chapter

http://dx.doi.org/10.5772/intechopen.73297

Abstract

Traditionally, the evolution of flora and fauna on the Earth as well as the evolution of their physical and chemical environment are considered separately. At the same time, when considering the global evolutionary changes, it becomes clear that the evolution of all these components occurs in close relationship and that they together constitute a unified evolutionary process. Thus, we should talk about their co-evolution and that the whole biosphere is a united functional system. In this chapter, we briefly discuss some of the major "inventions" of ancient life that are responsible for global biosphere transformations and which "worked" in the biosphere until now (photosynthesis, eukaryotic cell, multicellular organism, and the other findings). The evolution of the Precambrian life as well as the Phanerozoic stage of the biosphere evolution are considered in this context.

Keywords: evolution, co-evolution, ecosystems, biosphere, inventions of life

1. Introduction

Traditionally, when discussing biological evolution, specialists build independent evolutionary trees for different taxa, such as animals and plants, apart from the fact that many evolutionary events in each case would not have been possible without the concerted events in several macrotaxa. The evolution of life on the planet is the evolution of the entire biosphere as a united system [1–3]. For example, the emergence and development of higher flowering plants is interconnected with the development of pollinating insects. Classic examples of an ancient co-evolutionary relationship are lichens, which are a result of exo-symbiosis (mutualism type) of fungi mycelium and algae cells. More recent examples of symbiotic relationships are some species of ants and aphids, which are so deep that both cannot exist separately [4].

Moreover, the evolution of life on the planet occurs simultaneously with the transformation of biogenic habitat. Thus, in the geological timescale, it makes sense to consider the evolution of the biosphere as an integrated system, which includes not only the biota but also bio-inert and inert matter in terms of Vernadsky [1], as well as ecosystem, landscape, and the whole global planetary structure. Life as an active and most dynamic component of the biosphere permeates the entire environment. It gradually transforms environment and thereby changes the conditions of its own existence. However, bio-inert and inert matter, geological and landscape structure of the upper layer of the Earth's crust with a thickness of about 10 km (i.e., the biosphere) is much more inert.

Life with its activity quickly evolves and adjusts to the environment. Thus, there is a large temporal asymmetry between the processes of evolution of living and inert matter. The evolution of life changes in two ways: by spontaneous self-modification (genetic, epigenetic) and through accumulation of changes in the environment, partly caused by the activity of organisms themselves (ecological inheritance [5–7]). From this standpoint, the evolution of the biosphere consists of a sequence of leaps between qualitatively different stages. Each stage is related with some principally new "invention" of life, has a relatively long history of its own specific organization of its biogeochemical cycles, energy flows, and associated specific ecological relationships: producer (producents)-consumer (consuments)-reducer (reducents).

Small disturbances in balance of each such system caused by small changes of environmental conditions were usually compensated, thanks to adaptive evolution at the species level (appearance and disappearance of some species). At significant violations of the biogeochemical cycles, caused by external and/or internal reasons, global environmental crises took place, associated with mass extinctions of the traditional biological forms, replacing them with fundamentally new ones, the emergence and/or the wide dissemination of which was previously limited by competition.

It is easy to see that classical evolutionary phenomena of aromorphoses and ideoadaptations [8] are observed also at the biosphere level and almost simultaneously in different taxa, indicating the influence on the course of processes of co-evolution mechanisms. For example, the so-called great Permian extinction (end of Permian period, about 250 Mya, when more than 70% of species disappeared) dramatically decreased the total biological diversity [9]. This opened opportunities for the emergence of new forms, primarily in the Triassic.

As a rule, a number of forms of the earlier system do not disappear completely during crises. They are included in new ecosystems as representatives of evolutionary descendants of the more ancient biota. Moreover, despite significant evolutionary change, they usually are conserving successful evolutionary inventions of the past, and these findings allow them to "fit" into the new system of the biosphere. Modern biosphere utilizes many findings of ancient life and even could not exist without them. Among the main evolutionary innovations, which once established have not disappeared, and are used by living organisms of the contemporary biosphere, is the emergence of eukaryotic cells, which opened the road to evolution of multicellular organisms.

There are different ways of using the innovations of the past. One of the most striking examples is the discovery of the process of photosynthesis, which allowed terrestrial life to get out of the clutches of material resources, and the energy deficit of archaic biosphere that existed, apparently due to chemosynthesis. Photosynthesis gave the biosphere a huge gain in energy resources.

In order to be preserved, ancient life also changed its habitat. This is particularly noticeable in the case of bacterial forms of life. The traces of ancient biospheres surround us, and their numerous representatives are necessary for the existence of the modern biosphere, for example, the bacterial environment of the rumen of ruminants, where cellulolytic and other bacteria realize chemical transformations based on anaerobic processes—fermentation and decay of plant food components. The digestive system of mammals would be unable to digest plant food without help of endosymbionts—bacterial communities [10]. Thus, organisms with anaerobic metabolism, inhabiting once, apparently, the entire ancient biosphere, now exist in the form of microcosm in organisms of modern animals. Obviously, the evolution of the biosphere is reflected in mutually agreed evolutionary transformations of all life forms. Therefore, speaking about evolution of the biosphere, we should talk about biological co-evolution of all living on the planet. Important aspects—biochemical and physiological—of this process are studied in the framework of evolutionary physiology and biochemistry. However, since the co-evolutionary aspect is usually not discussed at all, many factors and evolutionary mechanisms often attract the attention of researchers who used to work at the organism and population levels.

The aim of this chapter is not to give a coherent and complete account of the events in the biosphere evolution (it is only because of the enormous amount of material that could be included in consideration) but touch only some general trends and some striking evolutionary events, reflected in contemporary forms of life. Our goal is to attract attention to biosphere aspect of the biological evolution [11], which often escapes researchers' notice. We will try to trace some of the most important innovations implemented in the process of evolution of life on Earth. These innovations, actively used by living organisms in both the past and the modern biosphere, were key endogenous factors of biosphere evolution as a whole system.

2. The early stages of the evolution of life

We know very little about the earliest stages of evolution of life on our planet. Most researchers believe that the life on the Earth appeared in the range of 3–4 Gya and was represented by prokaryotic organisms—Archaebacteria (at present allocated also another macrotaxa Archaea, which exists today, and from which, apparently, originated more complex prokaryotic organisms—[12, 13]). There is no doubt that by the end of the Archean (about 2.5 Gya), life was already widespread in the world, mainly in the aquatic environment. The oldest Archean prokaryotic biosphere probably used mineral resources available in the geospheres of the planet, and, in the first place, abiotic organic matter which is a non-renewable resource. Most likely, resources for this life were hydrocarbons that were typical for the crust and surfaces of many small planets in the solar system during its formation and during the early stages of evolution [14, 15]. As for the atmosphere, there is no consensus among geologists on this issue, and specialists suggest that it could be either predominantly methane, ammonia, or a mixture of carbon dioxide and nitrogen with a substantial admixture of methane and ammonia [16]. These hydrocarbons, especially liquid fractions, are possibly related to the so-called Archean and Proterozoic oil, which is poorly suitable for industrial use and has presumably abiotic origin. Sun has contributed to the existence of early life on Earth indirectly, mainly by supporting the chemical cycles and the overall temperature and climate balance of the planet.

About 2 Gya, there was a very important event which predetermined the direction of further evolution of life on the Earth: appearance, rise, and wide dissemination of photosynthetic organisms—cyanobacteria. With the appearance of a significant amount of atmospheric oxygen, ancient biota fundamentally changed because oxygen is poison for most anaerobic organisms, typical of the early Archean. There was the so-called Oxygen Catastrophe [17, 18]. Another—though perhaps no less important factor that contributed to a very rapid change of biota after the emergence of photosynthesis—was probably the exhaustion of available resources, that is, organic matter for the use by anaerobic organisms. You can imagine a variety of scenarios, up to the fantastic (e.g. burning out of hydrocarbons in an oxygen atmosphere) but there is no evidence in favor of any mechanism of the transition to the photosynthetic biosphere.

The result of the emergence and development of the first communities of photosynthetic and aerobic organisms was that biota has changed dramatically, and life and its evolution began to be related not only to the availability of resources of organic matter but also with the solar activity in the diapason of visible spectrum. With certainty, we can say that at this time already existed ecosystem biosphere organization with closed cycle of organic matter, which ceased to be a non-renewable resource. In addition to the phytoplankton in the ocean lived zooplankton [17, 18] and bacterial organisms. It is not difficult to conclude from the previous statement, it is this circular organization (i.e., producer-consumer-reducer) that defines and identifies up to the present time the main features of all forms of co-evolution of the biosphere life. We would also like to draw attention to the fact that, in this case, the organic matter of biotic origin, worked out by the organisms that preceded this "aerobic revolution," as well as atmospheric carbon dioxide have not disappeared and entered into a new metabolic cycle of the biosphere.

Approximately 1 billion years ago, in the life of the biosphere, there was another very important event: eukaryotic organisms appeared and after the phytoplankton crisis at the turn of Vendian and Riphean (about 800 Mya) came to dominate [18]. According to popular conception of symbiogenesis [19–21], they arose as a result of the merger of prokaryotic cells, one of which was to carry out functions of the nucleus. But the most important for the further evolution of life at the same time was the fact that eukaryotes were able to give rise to multicellular organisms. In contrast to the dense cell walls of prokaryotic cells, eukaryotic cell walls became more permeable, making possible a more intense and varied exchange of substances between organism and environment. In parallel, there was also a significant change in the whole system of cell receptors, necessary for the management of permeability, and principally new types of receptors appeared [22]. As a result, all this has contributed to origin of multicellular organisms (which was not possible in the case of prokaryotes) and accompanied with the structural and functional differentiation of individual cells or groups of cells. This was a necessary condition for the emergence of further tissues and organs with specific functions and features of their different structures. The emergence of multicellular organisms was, without doubt, a new revolution in the biosphere. By the beginning of the Cambrian (about 550–570 Mya), which is considered as the beginning of "the era of advanced life," the oceans were already home to many species of multicellular plants and invertebrate animals.

Exact period of coming organisms ashore is difficult to date because the accidental getting of organisms to coastal, small-scale and temporary existing reservoirs, where they could exist for some period, was not an exceptional event. In the upper Vendian (Ediacaria), in Cambrian, and even Ordovician periods reigned the so-called bacterial mats based on cyanobacteria communities. They covered shallow waters, filling intermittent irregularities of land. However, it is well established that in the Silurian period already existed "true" terrestrial multicellular plants with a complex structure and stem—psilophytes—and to the Carboniferous period the plants fully settled on land.

Further evolution of life on the planet, too, was uneven. This was largely due to the geological events and the impact of external factors of astrophysical nature on the planet, leading to crises and radical change of biota. The most striking well-known crises of this kind were the extinction in the Permian and Cretaceous periods, the beginning of which was characterized by a decrease of carbon dioxide concentration in the atmosphere and planetary cooling [23, 14, 15]. However, the dependence of all life and its evolution on photosynthetic organisms within the Phanerozoic history (i.e., ~570 million years and later) has never decreased, and the biosphere actually was and continues to be the united system that supports its own existence and operation of its components, using solar energy.

3. Phanerozoic eon

Phanerozoic eon of the biosphere evolution, which began with the Cambrian period about 570 Mya and is still ongoing, can be described as the time when life came out from the oceans and confidently conquered the land, which saw the emergence of all the modern types of plants and animals, and at last, man appeared, who became by the figurative expression of V.I. Vernadsky's [1] "geological force." The evolution of life in the Phanerozoic was uneven, as reflected in the fossil and geological records.

The beginning of the Phanerozoic eon was marked by the separation of the paths of the evolution of multicellular animals and dividing them into two large groups: invertebrates and vertebrates. This was due to the emergence of the first chordates (usually mentioned lancetnik that appeared in the lower Cambrian), which gave rise to all the past and present vertebrates, including humans).

As for plants, new groups of multicellular organisms emerged: first lower and then higher vascular organisms, for example, horsetails and ferns, which became a part of a major share of large-sized ground vegetation formed by Carbon. A little earlier, during the Carboniferous period, the ancestors of the first gymnosperms appeared in the upper Devon, and widely spread across the planet just after the Permian crisis, that is, in the Triassic. After the mid-Mesozoic eon, in early Cretaceous period (135–140 Mya), finally, flowering plants appeared, which were pollinated by pollinators insects.

It is hardly appropriate to discuss in this chapter many evolutionary and co-evolutionary events in the evolution of life in the Phanerozoic eon. Many special works are dedicated to the study of such events, including "a skeletal revolution," the emergence of systems of transport of substances inside the body, the "invention" of hemoglobin and blood circulation, the occurrence of thermoregulation of bodies, terrestrial respiration, senses, nervous system, brain, social behavior, and more [24]. From our point of view, the examination of the evolution of the entire biosphere is conveniently conducted in the framework of the so-called energy approach, especially as many evolutionary innovations are related with the global tendency of increasing in time of the energy flow through biosphere. Many researchers (e.g., see [14, 15, 25]) emphasized the importance of the energy characteristics in considering such systems as the ecosystems and biosphere as a whole.

4. The evolution of the biosphere as a united system of planetary life

Based on the earlier view, and also based on the fact that the evolution of life on the Earth should be seen as the *panbiosphere global process* of permanent mutual (reciprocal) adaptation of organisms in ecosystems and biogeocoenoses,[1] the *physico-ecological concept of biosphere evolution* has been developed [14, 15]. In the framework of this concept, the relationships between the various evolutionary processes at different levels of biological organization, as well as general problems of evolution of biosphere, are discussed. Observed during the Phanerozoic eon, increase of energy flow, passing through the biosphere, is understood as the physical evolution of the biosphere. This increase is a result of the process of the emergence of new plant producers and plant communities, which more efficiently used the Sun's energy. The rise of energy flow in its turn leads to increase in the production of organic matter used by consumers and decomposers, strengthening biosphere circulation and other important changes in planetary life.

In the early stages of the evolution of mechanisms for the use of solar energy, which began in the Proterozoic era, physical evolution can be correlated with the improvement of the chemical processes of photosynthesis, increasing the efficiency of chlorophylls. As a result, the evolution worked out the most efficient chlorophylls, used by plants up to the present time. Almost all of the main producers of organic matter in the biosphere, including the most important present producers, such as the highest vascular plants, in varying degrees, rely on sets of chlorophylls ***a*** and ***b*** [26]. Evolution of chemical aspects of photosynthesis completed to the Phanerozoic eon and elaborated optimal for Earth's conditions chlorophylls. Eukaryotic life, appeared in the Proterozoic began to go out on the land [18, 26, 28]. This, as already mentioned, was provided by the unique abilities of eukaryotes to create complex

[1]Biogeocoenosis is a large ecosystem considered within a framework of concrete plant communities [27].

multicellular organisms with different functions of specific tissues and organs, which gave a huge number of new possibilities and ways of existence of these organisms in the environment. For example, the extraction of water from under the top layer of soil by roots and carrying out of photosynthesis by organs above the surface, where lighting conditions are usually better—one of the abilities unavailable to single-celled plant organisms.

The growth of the photosynthetic ability of terrestrial plant communities and coupled with this process of physical evolution of the biosphere occurred at this stage due to the occurrence and development of adaptations, which led to the increase of the area of photosynthetic surface of leaves and sometimes other formations. Thus, Phanerozoic evolution of the biosphere has largely been associated with morphological changes [18, 28] of terrestrial plants and, respectively, of multicellular animals—consumers of organic matter and the oxygen which these plants produce.

It is also important to remember one more factor that played a huge role in the evolution of life on the planet during the Phanerozoic era, namely, the presence of an oxygen atmosphere [17, 18, 29]. It was oxygen that offered animals to intensify metabolism, to move from water to land, and significantly increase the size of some of them. The concentration of oxygen in the atmosphere in the Phanerozoic eon fluctuated, and in the epoch of coal accumulation exceeded, apparently, the contemporary one. Here, we return to the thesis about the ecosystem organization of the biosphere, representing the *over-organismal living system* [3, 7, 30], sometimes figuratively called the "super-organism."

The importance of animal components of the biosphere is beyond doubt: animals in combination with plant and reducents organize a closed and well-functioning biosphere cycle; in other words, a continuous process of "biosphere metabolism" [30]. Evidently, evolutionary changes in plant and animal species are linked also via the change of geospheres, including the atmosphere.

Here, we do not consider the mechanism of physical evolution of the biosphere, which is described in the monograph [14, 15, 17] and is based on the well-known notion that the biosphere is a united functional system. Note, however, that in the framework of this approach, the physical evolution is explained as a consequence primarily of external astrophysical factors. These are the factors of the scale of the galaxy, influencing the intensity of geological processes and outgassing of carbon dioxide from the bowels of the Earth with a period of the galactic year of about 200 million years and factors of the scale of the solar system, causing periodic oscillations of orbital parameters of our planet and climatic changes on the Earth every few tens of thousands of years. The impact of these factors stimulates evolutionary processes in the biosphere and leads to an increase in the flow of energy through it, as a result of the selection and wide dissemination of producers, which more effectively are using the flow of solar energy (**Figure 1**).

In the end, during the Phanerozoic eon, new macrotaxons of plants and animals and new ecosystems and biogeocoenoses arose, energy flow through terrestrial plant communities increased by nearly two orders of magnitude, and the speed of circulation of organic matter in the biosphere increased dramatically. All this affected the course of biological evolution [14, 15].

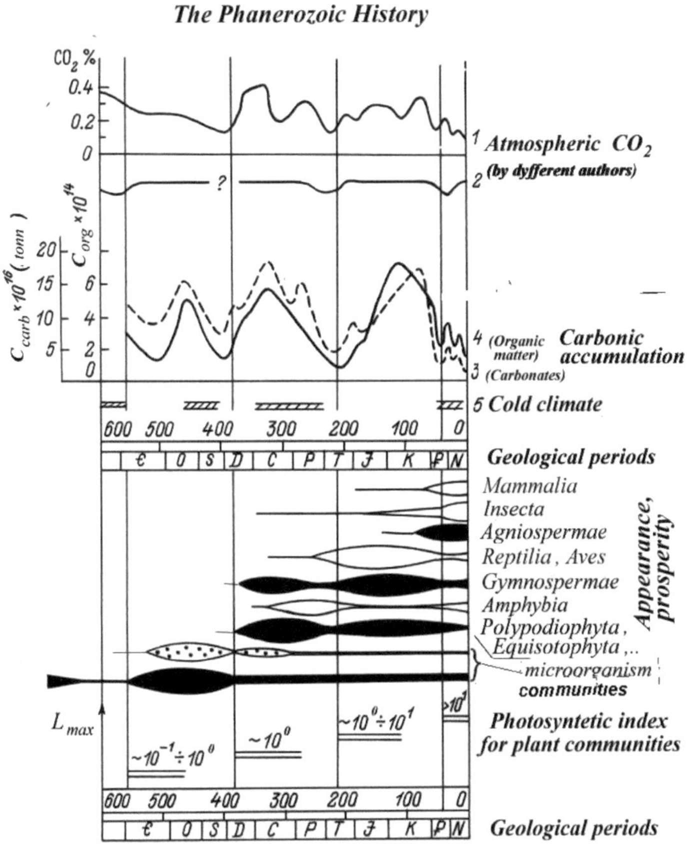

Figure 1. Phanerozoic evolution of the biosphere. The oscillations of carbon dioxide concentration in atmosphere, the epochs of global cooling, the development of different macrotaxa, and increase of photosynthetic index (it characterizes the community productivity) are given by [14, 15].

To conclude this section, we note that to date, the possibility of increasing the energy flow through the biosphere, ecosystems, by traditional ways, that is, by increasing the photosynthetic index of the plant communities, seems to have exhausted—changing the trends of evolution of life on Earth.

5. Some general features of the evolution of the biosphere

Laws of evolution of ecosystems and the biosphere, the known laws of micro and macroevolution, as well as regularities known from evolutionary biochemistry and physiology should be considered and coordinated together. One of the features of the evolutionary processes of the biosphere is that every time they take place in conditions of irreversible changes on the Earth, produced by previous evolutionary process. Therefore, we can speak of a kind of "memory" of the biosphere, and what is contained in this memory to a certain extent directs, canalizes occurring in the biosphere subsequent evolutionary changes. The most important

http://dx.doi.org/10.5772/intechopen.73297

elements of this memory are those which are fixed in the genomes of organisms and, consequently, the particularities of their morphology and biochemical mechanisms and is also reflected in the geospheres of the planet, its geological history. These, for example, are the emergence of an oxygen atmosphere about 2 Gya; the emergences of various biogenic sediments, rocks, and minerals in the bowels of the Earth; the emergence of modern water; and atmospheric balance of the planet. That life controls certain features of its own evolution is reflected in the paradigm of *autocanalization of the biosphere evolution* [30].

All this may suggest that the process of biosphere evolution is similar in some aspects to successive deployment, which resembles the ontogenetic development, when each phase starts only after completion of the previous one. However, this similarity is not complete though, because the biosphere changes are not pre-programmed. The similarity can be discussed only in relation to the evolution of the entire biosphere as a whole but not in relation to changes in ecosystems, populations, or genotypes. After all, the final goal of ontogenesis is the stage an adult organism reaches with certain morphological and physiological characteristics that allow it to exist and reproduce in a particular environment. But the biological and biosphere evolution, most likely, do not have any final goal. The only permanent "goal" (more correct to call it an imperative) of the functioning of the biosphere is the preservation of life itself through new adaptations and co-adaptations of different organisms and ecosystems in a changing environment. In other words, imperative for the biosphere as a system is the maintenance of life on the planet in any way, like using relatively small adaptive changes in the large components or (and) through significant evolutionary changes on the species level, even at the cost of destruction of entire taxa [15, 17, 18].

Observable predetermination and directedness of the physical evolution of the biosphere toward a more energetically developed and, accordingly, productive state with high biodiversity [31] do not mean predetermination of biological evolution. Selection usually acts on a limited number of parameters and fixes the first met biological forms by chance, as long as they meet these conditions. Therefore, features of the natural environment that occurs at a particular stage in the evolution of the biosphere, not rigidly determine the ways of biological evolution [17, 30, 32] does not require that the performer of a particular ecosystem functions were very specific, for example had concrete origin.

Thus, the reason for the directionality of biological evolution toward complication, leading to the emergence of increasingly complex biological forms (while retaining many archaic one), is the evolution of the whole biosphere [14, 15, 17]. As was mentioned above, in the process of this evolution takes place the increase of energy flow through biosphere, as well as the complication of the mechanisms, supporting conditions on the planet, favorable for more advanced life forms ("planetary homeostasis"). Other aspects of this process are the *complication of relationships between components of the biosphere and increase the number of simultaneously present in the biogeochemical cycle of organic matter.* All these phenomena involve irreversible changes of geospheres of the planet and accumulation of "burden of evolution" in the form of morphofunctional and morphogenetic changes of the organisms and, consequently, constraints on possible ways of further biological evolution [17, 30]. Therefore, we should talk about autocanalization of biosphere evolution, which is a consequence of the fixation of the irreversible changes occurring in the geosphere and the biosphere.

6. Some main trends of evolution of the biosphere

Let us summarize now the most important trends in the evolution of the biosphere, some of which have been briefly described earlier. First, it should be noted, that earthly life was very lucky, because even the most severe crises on the planet could not bring to extinction of all taxa or to drive biosphere back to earlier developmental stages. Throughout its evolution, the biosphere coped with the changes; moreover, they stimulated her to further evolution ("everything that does not kill me only gives me new strength"—Nietzsche). The presence of biosphere-geosphere memory and genetic memory of the organisms is allowed not to lose the successful discoveries, which is, in fact, technologies of survival. Thus, the process of evolution led to the emergence of increasingly complex forms.

The main features and trends of evolution of the biosphere are briefly outlined as follows:

- the increase in the flow of energy flowing through the biosphere (this conclusion was made on the basis of data for Phanerozoic eon);
- an increasing amount of organic matter simultaneously involved in the biotic cycle (reliably known for the Phanerozoic eon);
- ecosystem structure of the biosphere (at least from mid-Proterozoic eon);
- increase in the completeness of biogeochemical cycle of the biosphere (from one-directed flow in open ecosystems of bacterial mats up to ~95% completeness of cycle in a number of ecosystems by the end of the Phanerozoic eon);
- expansion of the scope of buried bio-inert substance of the planet (from Archean);
- complication in the structural and functional diversity of the biosphere and its components (differentiation) that was required for the survival of many organisms to have more complex behavior and, ultimately, contributed to the emergence and development of the nervous system and brain (Phanerozoic eon);
- increasing biodiversity [31] and diversity of ecosystem types (at least from the Proterozoic eon);
- increase in the sustainability of the biosphere and its components in the case of non-catastrophic disturbances is related to the increase in biodiversity.

All of the above and a number of other evolutionary changes occurred under the pressure of autocanalization mechanisms, which did not determine strictly each step of biological evolution but, as in the case of ontogeny, implied continuity. This is related to some analogy of the processes of development and evolution, allowing to call the evolution of the biosphere by "non-directed ontogeny," in which concrete specific forms of organisms are not strongly predefined, but the general characteristics of interaction of the whole system of the biosphere with the inanimate nature are more predefined. At least on the Earth, this "ontogenesis" led to the emergence of increasingly complex biological forms right up to the human beings and complication of the system of the biosphere. *The main functional imperative of the biosphere at all stages of its evolution was the preservation of life itself on the planet in any way.*

We do not dwell here on the philosophical premises and conclusions, following from this assertion. Much on this subject can be found in the monograph by O. Bazaluk [33]. Let's look at the process of evolution not from the point of view of changes in the structural and functional features of living things but as the process of creation of new and existing means of survival and exploitation of the environment by life. From this point of view, the evolution of the biosphere is the invention, preservation, improvement, and accumulation of different technologies of storage, and the implementation of process of life. With these positions, in the evolution of life on the Earth were the following main stages:

- the appearance of protobionts with memory to store information about the successful technologies of survival. The main innovation of this phase—the emergence of the genetic code (apparently, early Archean eon);
- the emergence of technologies for the use of mineral resources of the planet. The main innovation of this phase—the emergence of chemosynthesis and of the primitive metabolism (around the same time as the previous stage);
- the emergence and rise of organisms able to use photosynthesis. The main innovation of this phase is the development of mechanisms of photosynthesis to use energy from the Sun. In parallel, the emergence and spread of aerobes and the development of aerobic metabolism; in fact, fixation of the ecosystem organization of planetary life (upper Archean eon, Proterozoic eon) [34];
- the emergence of eukaryotic cells as believed to be the result of symbiosis of prokaryotes [19–21]. The main innovation of this stage—effective new nuclear apparatus of eukaryotic cells and the development of specific method of transfer of genetic information—sexual reproduction. Besides that, there was altered cell membrane, a complex of new cell receptors emerged, allowing better control of the permeability to various chemicals (middle Proterozoic eon);
- the emergence and broad spreading of multicellular organisms. The main innovation of this phase is the emergence of the communication system between cells and the mechanism of their differentiation in the ontogeny of multicellular organisms (upper Proterozoic eon).

It is important to note that with the appearance of another main innovation, the old evolutionary findings usually do not disappear. They not only continued to exist and develop but also combined with newly acquired features. Similar features of evolution can be seen not only in the early stages of the evolution of life but also later and also when considering the emergence of organs, physiological systems, and the development of physiological functions. When studying different groups of organisms, a number of particular regularities have been opened to confirm these general regularities (K. Baer, G. De beer, R. Garstang, E. Haeckel, E. Cop, S.V. Meyen, A.N. Severtsov, A.L. Takhtajan, K. Waddington and I.I. Schmalhausen, etc.). These general regularities, apparently, are also applicable to ecosystems and the biosphere [14, 17]. Close in its meaning regularities were also aptly formulated by L.A. Orbeli in the form of "principles of evolutionary physiology" [35]. Among them are the following:

- the principle of intensification of the processes that provide the functionality of biological systems (including in terms of resource consumption);
- the principle of the increasing of multifunctionality of the components of biological systems;
- the principle of the increasing of duplication components of biological systems that perform a particular function;
- the principle of overbuilding (or superstructure): new functions do not simply replace the old one but are built on the old functions, replace them, and manage them.

It is not hard to see that these principles which were formulated for evolutionary physiology can explain events in the evolutionary history of the biosphere. Apparently, they are universal and in varying degrees applicable to explain the evolution of biological systems of various levels of the organization.

Homo sapiens, appeared at the end of Phanerozoic eon, began to use for their needs fossil raw materials and energy sources inaccessible to other organisms (e.g. fire, nuclear energy). They continued general tendency of intensification of energy flow and biogenous cycles in the biosphere evolution. Humans' extremely rapid spreading across the planet and increasing role in biosphere processes is obliged first of all to the fact that they use language for communication and transmission of information about their technological discoveries. Biological evolution has ceased to play a crucial role in their survival on Earth.

7. Conclusion

This chapter is not an exhaustive description of the evolutionary events in the biosphere. However, we hope that we were able to show the resemblance and the connection of the evolution of biological systems at different levels of biological organization in the process of development of life on Earth.

We discussed that the evolution of the biosphere is uneven, that is, periods of abrupt change (crises) is changed by much longer periods of relative stability. What could be the cause of a sharp, "explosive," and at the same time, global changes in the entire biosphere? Based on the earlier description, we can distinguish two types of reasons, conventionally, dividing them into "external" and "internal." The first type includes global change of the abiotic environment that occurred under the influence of planetary and cosmic factors. The second type which consists of the "main innovation" is a radical change in the uses of the environment associated with the emergence of new technologies of survival, including new ways of organizations and associations of living systems (eukaryotes, multicellular organisms, ecosystems, etc.).

In conclusion, it is important to say, that humankind, which, according to V.I.Vernadsky [2] is the new "geological force", plays an increasingly important role in key biosphere processes. Now, we can add, that humankind is the force, which became a new factor of evolution, realizing conception of autocanalization of evolutionary process. However, despite the rapid development of non-biological technology, man, being a biological creature, cannot ignore its own earthly nature, because all the main stages of biosphere evolution are reflected in the human genetic memory and in the processes of his development. That is why existence of man is possible only as co-existence with all other living creatures [6]. This need in the "other" is applied to all organisms and was "imprinted" early in evolutionary history, before the emergence of the mind, and originates probably from Archean times, when the first simplest ecosystems appeared. After all, to find a "friend" and assistant contributed to the survival. This need, combined with the pursuit of invention of new technologies, creates one of the main contradictions of the inner world of modern man. Here, we are coming close to the non-biological evolution of the noospheric environment [2] in which the human beings, the inventors play an important role and have to take responsibility for the destiny of his beautiful home—the biosphere. Whether their inventions are accepted depends largely on the culture in which they exist.

Author details

Vladimir F. Levchenko[1]*, Alexander B. Kazansky[1,2] and Marat A. Sabirov[1]

*Address all correspondence to: lew@lew.spb.org

1 Sechenov Institute of Evolutionary Physiology and Biochemistry of the Russian Academy of Sciences, St. Petersburg, Russian Federation

2 Federal Almazov North-West Medical Research Centre, St. Petersburg, Russian Federation

References

[1] Vernadsky VI. The Biosphere (in Russian); 1926. Transl. Langmuir DB. Mark AS, editors. McMenamin, New York: Copernicus; 1997. 192 p

[2] Vernadsky VI. The biosphere and the noosphere. American Scientist. 1945;**33**:1-12

[3] Lovelock J.E. Gaia: The Practical Science of Planetary Medicine. Gaia Book Ltd, 1991. 192 p

[4] Shaposhnikov GH. Directedness of evolution. Zhurnal Obshcheĭ Biologii. 1977;**38**(5): 649-655

[5] Odling-Smee JJ, Laland KN, Feldman MW. Niche Construction: The Neglected Process in Evolution. New Jersey: Princeton University Press; 2003

[6] Kazansky AB. Bootstrapping of life through Holonomy and self-modification. In: Dubois DM, editor. Proceedings of the 9th International Conference on Computing Anticipatory Systems. Vol. 1303. Melville, NewYork: American Institute of Physics; 2010. pp. 297-306

[7] Kazansky AB. Planetary bootstrap: A prelude to biosphere phenomenology. In: Dubois DM, editor. Proceedings of the 6th International Conference on Computing Anticipatory Systems. Melville, New York: American Institute of Physics; 2004. pp. 445-450

[8] Severtsov AN. Essays on the Theory of Evolution: Individual Development and Evolution. Berlin: Publishing House RSFSR; 1921. 309 p

[9] Zherichin VV, Rautian AS. Phylocoenogenesis and evolutionary crises (in Russian). Zhurnal Obshcheĭ Biologii. 1997;**58**(4):20-47

[10] Golikov AN, Bazanova NU, Karabekov ZK. Physiology of Farm Animals (in Russian). Agropromizdat: Moscow; 1991. 432 p

[11] Starobogatov YI, Levchenko VF. Ecocentric concept of macroevolution (in Russian). Zhurnal Obshcheĭ Biologii. 1993;**4**:389-407

[12] Schopf J. Fossil evidence of Archaean life. Philosophical Transactions of the Royal Society of London. Series B, Biological Sciences. 2006;**361**(1470):869-885

[13] Vidal G. The oldest eukaryotic cells (in Russian). V Mire Nauki. 1984;**4**:14-24

[14] Levchenko VF. The three Stages of the Evolution of Life on Earth (Monograph in Russian). Germany: Lambert Academic Publishing; 2011

[15] Levchenko VF. The Biosphere: The Stages of Life (the Evolution of the Parts and the Whole) (Monograph in Russian). ISVOE: Saint-Petersburg; 2012. 264 p

[16] Kenyon DH, Steinman G. Biochemical Predetermination. New York, London: McGraw-Hill Book Company; 1969

[17] Levchenko VF, Kazansky AB, Sabirov MA, Semenova EM. Early Biosphere: Origin and Evolution. In: Ishwaran N, editor. The Biosphere. InTech; 2012

[18] Fedonkin MA. Two Annals of Life: Comparison Experience (a Paleobiology and Genomics about Early Stages of Evolution of Biosphere) (in Russian). Geoprint: Syktyvkar; 2006

[19] Kozo-Polyansky BM. A New Principle of Evolution. Essay on the theory of symbiogenesis (in Russian). Moscow; 1924. Transl. by Fet V, Fet V, Margulis L. editors. Cambridge, Mass.: Harvard University Press; 2010

[20] Margulis L. Symbiogenesis. A new principle of evolution rediscovery of Boris Mikhaylovich Kozo-Polyansky (1890-1957). Paleontological Journal. 2010;**44**:1525-1539

[21] Margulis L. Symbiosis in Cell Evolution. 2nd ed. New York: Freeman WH; 1993. 452 p

[22] Shpakov AO, Pertseva MN. Signaling systems of lower eukaryotes and their evolution. International Review of Cell and Molecular Biology. 2008;**269**:151-282

[23] Yablokov AV, Levchenko VF, Kerzhentsev AS. Sketches of biospherology. 1. There is a solution: Go to the controlled evolution of the biosphere. Philosophy and Cosmology. 2015;**14**:91-117

[24] Svidersky VL. Modern problems of evolutionary physiology. In: Darwinism: History and our Time. Leningrad: Nauka; 1988. pp. 202-210

[25] Odum PE. Energy flow in ecosystems: A historical review. American Zoologist. 1968; **8**(1):11-18

[26] Meyen SV. Fundamentals of Paleobotany (in Russian). Moscow: Nauka; 1987. 228 p

[27] Sukachev VN. Fundamentals of Forest Typology and the Biogeocoenology (in Russian). FAV Works. Vol. 1. Leningrad: Nauka; 1972. 424 p

[28] Zherikhin VV. Selected Works in Paleoecology and Philocoenogenomics (in Russian). Moscow: KMK; 2003. 542 p

[29] Levchenko VF. Evolution of Biosphere before and after the Appearance of Man (Monograph in Russian). Saint-Petersburg: Nauka; 2004. 166 p

[30] Levchenko VF, Starobogatov YI. Self-regulatory evolution of the biosphere: The dynamics of the diversity of the organic world in time and space (in Russian). In: Materials 40-Th UPR Session. Saint-Petersburg: VSEGEI; 1994. pp. 30-32

[31] Brodsky AK. Biodiversity: structure, problems, and perspectives of conservation. Archives of Zoological Museum of Moskow State University. 2016;**54**:380-396

[32] Levchenko VF, Starobogatov YI. Succession changes and evolution of ecosystem (some problems of evolutionary ecology) (in Russian). Zhurnal Obshcheĭ Biologii. 1990;**5**:619-631

[33] Bazaluk OA. The Theory of Evolution: From the Cosmic Vacuum to Neural Ensembles and in the Future (in Russian). ISPC: Kiev; 2015. 312 p

[34] Kovalev OV, Kazansky AB. Two Faces of Gaia. University of East London: Gaia Circular; 1999;**2**(2):9

[35] Orbeli LA. The main problems and methods of evolutionary physiology (in Russian). In: Evolutionary Physiology, Part 1. Leningrad: Nauka; 1979. pp. 12-23

Chapter 10

Development of Multicellularity: Social/Economic Aspects

Gennady A. Savostyanov

Additional information is available at the end of the chapter

http://dx.doi.org/10.5772/intechopen.73658

Abstract

This article describes a philosophy of an arising multicellularity on the basis of division of functions between cells. Laws of the division are discussed in elementary multicellularity units, called histions. There is a variety of the histions that have different social structures. Several parameters have been proposed to describe them quantitatively and to systematize them by means of a periodic table. Consideration is given to the rules that govern polymerization of histions as well as the formation of regular cellular networks using them. It is shown that these types of networks could serve as biological tissue models that enable one to predict the tissue development. It has been found that arising multicellularity can result in a drastically decreased metabolites production per cell and thus creates the need in their economically justified unequal distribution.

Keywords: evolution physiology, division of functions, multicellularity, cell sociology, measuring of development, cell network, spatial organization of tissue, cell economic

1. Introduction

It was found long ago that the cell specialization and integration (i.e., the division of "labor" between cells and/or the cells cooperation) form the basis for a multicellular organism development. On that ground, Virchow considered an organism as a cellular state and emphasized its social aspect. Later on, Chandebois [1] and Gass and Hall [2] discussed cellular sociology. Then, with the advances in genomics and proteomics, scientists [3] discussed molecular sociology. The problem of labor division was relegated to the background in subsequent theoretical researches which focused on the ensuing altruism and on its benefits [4–8]. Some experimental studies were devoted to specific mechanisms of the cell specialization/integration [9–12]. Even bioeconomics was suggested for studying economic aspects of development [13, 14].

But both social and economic aspects of the multicellular organism development due to the labor division have been understood rather intuitively and/or qualitatively, remaining unexplored so far. Therefore, many significant development features are still incomprehensible. For example, what laws of the labor division govern the organism development? Why does the development have such characteristic features as cycling, directivity, reiteration, and parallelism? How are potencies gained and realized? What kind of parameters could be used to measure a degree of development and to predict those results? How we could build mathematical models of natural systems in the form of a periodic table? Any answers to these questions are unavailable today. Accordingly, there is not any theory that would be able to measure and to predict development of organisms as well as to construct their non-genealogic natural systems.

We think that a formal description of labor division could be used as the basis for analysis of social/economic aspects of development and for the successful elaboration of their predictive theory, based on an abstract model of the elementary multicellularity unit as an example. It is common knowledge that a major breakthrough in the understanding of the phenomena under study often began from building those abstract models. The simplest cellular groups, which arise as the result of the functional division between cells, could be such kind of models, as applied to multicellular organisms, and these groups will be elementary multicellularity units. In our case, they are called histions [15]. To describe a histion development, we propose a formal language whereby some new encouraging results have been obtained [16, 17]. They are summarized below.

2. Basic concepts

Our work uses the following basic notions:

- list L of functions a, b, c, d,… to be divided;
- list of performers (cells designated as circles);
- potencies to divide functions;
- potency realization conditions; and
- the organism as a performer who executes the entire list L.

The basic concepts are outlined below:

List L of functions are to be divided wherein the functions are lettered a, b, c, d …. An organism can perform a function from the list L in three different regimes, such as:

1. Autonomous survival (AS), which is inherent in single-cell organisms. In this regime, we denote function performers by circles and functions—by small letters above the circles (see **Figure 1**, line *m*, wherein the cell is in column 0). All functions of that sort are fulfilled for themselves only by means of the archaic processing. Thus, they lack any potency to the division of functions.

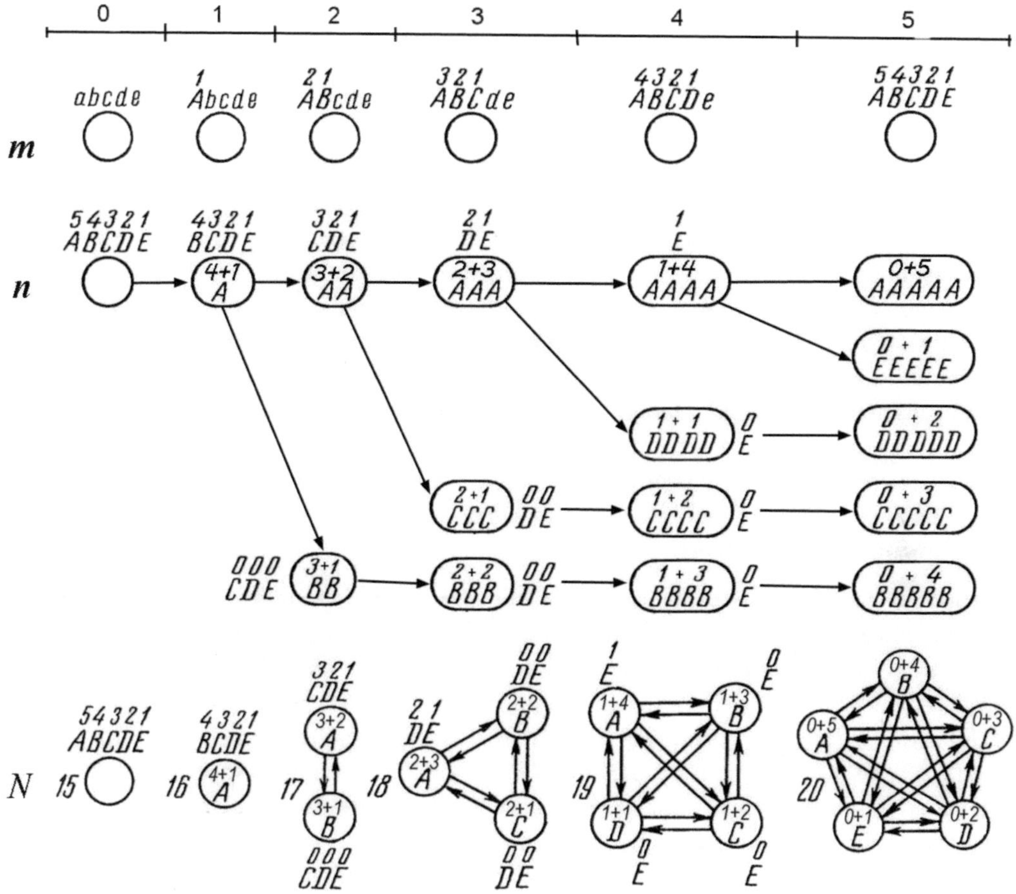

Figure 1. Schematic representation of elementary development acts. The upper line shows a growing number m of the functions that are gaining generative potencies and storing the sequence which is designated by numbers above capital letters. The middle line n illustrates realization of potencies based on the fact that the ancestor of the m = 5 line generates five daughter cells transforming them into more and more specialized ones. An increased number n of specialization acts leads to a rise in the cellular energidity (energeticity) which is denoted by the number of letters inside ovals. The lower line N schematically shows the cell specialization/integration stages and the formation of the simplest histions. The ciphers before histions indicate the total number N of their development acts.

2. Acquisition of potencies (AP): In this regime, some novelties are carried out which impart the division and specialization potencies to the functions, with the improved processing of them. In this case, we denote functions by capital letters above the circles and potencies—by numbers above the letters (**Figure 1** *m*). The number m of functions transferred into the AP regime is a significant parameter of development. A sequence of the transfer is fixed by the number of the gained potencies indicating a phylogenetic age of the function. These numbers represent an arithmetic progression of integers 1, 2, 3, …, *m*. The novelties lead to the appearance of an ancestor with *generative* potencies who is capable of giving a different-type offspring that is suitable to specialize in various functions.

3. Realization of potencies (RP): In this regime, an ancestor produces m daughter cells using a series of asymmetric mitotic cell divisions accompanied by transferring his own potencies to them (**Figure 1** *n*).

The arising daughter cells convert the obtained potencies into structural ones and become committed (or determined). The determined functions as well as those potencies are designated both by capital letters and by numbers within the circles (**Figure 1** *n* wherein the left and the right numbers designate non-realized and realized functions, respectively). The number n of the committed cell functions, that have obtained structural potencies, is the second significant parameter of development. It also defines the cell specialization degree varying within the range $0 \leq n \leq m$. The committed cells can realize their potencies via symmetric mitosis with the similar-to-themselves cells production as well as through specialization and improved processing of the function fulfillment. In this case, the number of function specialization acts corresponds to the number of the function potencies and appears to be variable. The same degree of the achieved specialization is provided for all functions by the fact that young functions get to specialize from a more and more differentiated state and perform the specialization with increasing numbers of partners.

The arising specialized cells integrate into histions by way of the metabolic cooperation and by the service exchange as well (**Figure 1**, *N*). Histions are elementary social units of multicellularity. Each change of function fulfillment conditions is the elementary act of histion development. Transitions AS → AP create generative functional potencies for the production of different-type specialists and well correlate with anagenesis. Transitions AP → RP convert these potencies into the structural ones and realize them *through* the cell specialization to correlate with cladogenesis. The cell integration as well as the composition and structure of histions are conveniently modeled by graphs, as shown in **Figure 1**. Therein, cells and functions are designated by circles and letters, respectively, and intercellular communications by arrows.

3. Quantitative characteristic of histion potencies

The potencies gained can be evaluated quantitatively. In this case, the basic function division parameters *m* and *n* not only permit the whole pool of potencies (the potenciom) to be found but also they make possible the division of these potencies into individual sorts. Besides, we may discuss the pool structure as well as evaluate pool changes in the process of development. The dynamics of potency variations has been analytically treated in [16, 17] and graphically shown in **Figure 2** (here values *m* and *n* were increased to 10 for clearness).

Clear regularities can be traced in tendencies to changing numerical relationships between different potencies for every *m* value. The intersection points of the curves that show the number of potencies divide a histion life cycle into five intervals. So, a projection of the first point of intersection on to abscissa (shown by the left broken line) limits a histion "childhood" period. The next point of intersection (shown by the left solid line) restricts a histion "juvenile" period. The interval between the two solid lines corresponds to the histion "youth" and is characterized by a maximum of non-realized structural potencies and, accordingly, by the best possibility for adaptation. The point of intersection of non-realized generative potencies

http://dx.doi.org/10.5772/intechopen.73658

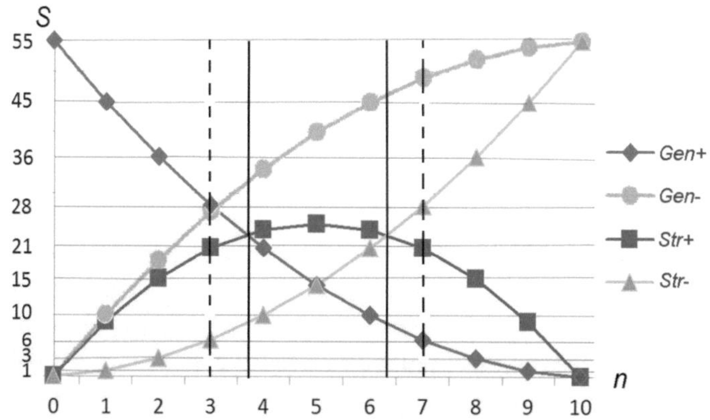

Figure 2. Variation of the number of generative and structural potencies in developing the m = 10 histion. Here the quantity n is laid off as abscissa and the number S of potencies as the ordinate. Gen + and Gen − are the numbers of non-realized and/or realized generative potencies, respectively. Str + and Str − are the numbers of non-realized and/or realized structural potencies. The left broken line is drawn through the first point, thus limiting a "childhood" zone (i.e., a cycle phase, rich of generative potencies and poor of structural ones). The left solid line is drawn through the second point. Both lines restrict a "teen-age or juvenile" zone (a phase with the lesser number of generative potencies but with the increased number of structural ones). The third (central) point bisects the life cycle. The right solid line is drawn through the fourth point. Both solid lines bound a zone of the adaptive maximum or "youth" (a cycle phase with the maximum number of isotopes and nonrealized structural potencies, as well). A cycle zone laying between the right solid line and the broken line corresponds to the phase of the histion "maturity," wherein the number of realized structural potencies is high and a stock of non-realized ones is diminishing but remains therewith pretty high. In addition, there are the rest of non-realized generative potencies. Further to the right, a zone is situated that corresponds to the "old-age" phase when the histions have realized all potencies, thus losing their plasticity and environmental sensitivity, whereas their regeneration capacity and adaptability are minimized.

and realized structural potencies divides the life cycle in two. Spacing between the right solid line and the broken line corresponds to the histion "maturity." The "old age" and the end of the cycle lie farther to the right.

Thus, the explored dynamics of potencies for the first time puts forward quantitative criteria for the life cycle division into the five periods (that had long been understood by intuition only), such as: "the childhood," "juvenile" period, "youth," "maturity," and "old age" and marks those initial and end points.

4. Law of conservation of histion potencies

The above-described dynamics of potencies enables one to come to the following important conclusion: The total number *S* of all histion potencies is dictated in every histion family by a value of parameter *m* (or by the number of functions available in the AP state). Hence, *S* varies only as parameter *m* changes, remaining unchanged with *m* constant. This concept is valid as a governing law in application to the "labor division" procedure. We can now state the law as follows: In the development of histions, the total number of potencies remains constant at any value *n*, but the ratios between generative and structural potencies and/or between realized and non-realized potencies are variable. This law is graphically shown in **Figure 2**.

Two important consequences follow from this law: The first one is that the development is bound to be cyclic because it has its own start (a universal ancestor with non-realized potencies) and its end (when all of the potencies have been realized). The second consequence is that any histion development inevitably involves a division of cells into stem cells and non-stem cells for the following reason.

In any development cycle with constant *m,* the proportion of non-realized generative potencies decreases and the number of specialized histion cells grows. As a result, there comes a time when there appears a shortage of potencies for all cells. This occurs in the point of intersection of two curves: one corresponding to the number of realized potencies and the other to the number of non-realized generative potencies (**Figure 2**, left broken line). Beginning with the moment of the passage through that point, a share of the arising specialized cells (the functions of which will have generative potencies in the AP state) will decrease. In **Figure 1**, such functions are denoted by capital letters with digits above them. Cells having such functions will be the stem cells. Yet, most of the arising cells have functions that remain without generative potencies in the AP state. In **Figure 1**, they are denoted by capital letters with *0* above them. Cells of this sort become the working cells that are capable of generating alike descendants only. Therefore, the law of conservation of histion potencies for the first time offers a simple explanation of the cyclic development as well as of the causes for the division into stem cells and specialized cells.

5. Evaluation of the progressive histion development

An integral measure N of the histion progressive development is to be found using parameters m and n that can be tested experimentally. As such a measure, the total number of development acts, can be taken, whereas N will be composed of two terms, such as: a sum S of all function potencies and the number n of the specialized functions realizing structural potencies, as follows:

$$N = S + n, \text{ or } N = \tfrac{1}{2}(m + 1)m + n \quad (1)$$

For the histions in **Figure 1**, *N* is shown in front of each of them. This quantity can be easily determined from histion structures. It is also possible to solve the inverse problem, that is, to define both m and n and a structure of a histion from *N* value [15, 16].

I shall note that up to now, there has been no convenient parameter offered to evaluate the progressive development. We were fortunate to discover it owing to the formal description of the labor division only. It has been suggested to name the offered unit *(N)* of progressive development as *Lamark* [16, 17].

So, *m* and *n* parameters not only make possible a quantitative dynamics description of potencies but also they allow a measure to be found for evaluating their progressive development. However, our description is a theoretical one so far. Testing of the predictable dynamics of potencies shall now become an important challenge for experimental biology of development.

6. Development rules

Division of functions between cells is governed by a set of simple rules or postulates [15, 16]. Several examples of such rules are cited below:

1. Every histion cell is provided with a chance to fulfill any function from set L. Only function fulfillment conditions are variable.
2. Every cell can specialize for fulfillment of any single function only. In this case, n parameter will indicate the number of the cell types involved in the histion.
3. Histion cells are equal in the number of partners.
4. Integration of partners is always mutually beneficial and so on.

These rules may be modified to take into account different environmental conditions and to obtain histion families with varying compositions and social structures [16, 17].

7. Law of periodic histion development

By analyzing the above development rules, we have found that there exists such a combination of rules which can be characterized by a law of periodic histion development stating: The histion composition and structure repeat periodically as the total number *N* of development acts increases.

Thus, we can classify histions using a periodic table by arranging them in ascending order *N* and combining the same m-value histions in a row and the same n-value histions in a column, as shown in **Figure 3**. This table gives us an idea of a multitude of directions available for a cell specialization as well as of their relationship variants, that is, of the histion social structure.

The parameters of this table have a biological significance and are suitable for a quantitative analysis of progressive development. Thus, that row number m indicates the number of functions in the GP state while the column number n indicates the number of functions in the RP state. The number N that is cited in the upper left-hand corner of a table cell indicates the total number of histion development acts. In addition, every table cell contains the number H of isotopes which differ in their cell compositions. So, an isotopic coordinate appears and the table becomes three dimensional. The H value is equal to the number of combinations of m elements taken n at a time that is cited in the lower left-hand corner of a table cell. Within each table row, the number of isotopes gradually increases, then it reaches a maximum in the middle of the row (area corresponding to the adaptive maximum zone), and is minimized again at the end. That forms the basis for the divergence in the beginning of the development and for convergence—in the end. The table, which is formally based on the Pascal's triangle, serves as a natural parametric system of histions, demonstrating their megaevolutionary variability. A histion position in the table uniquely determines all the properties of a histion which is also inherent in natural systems [18]. The proposed table offers certain advantages over any options built on other, rather intuitive, grounds [19].

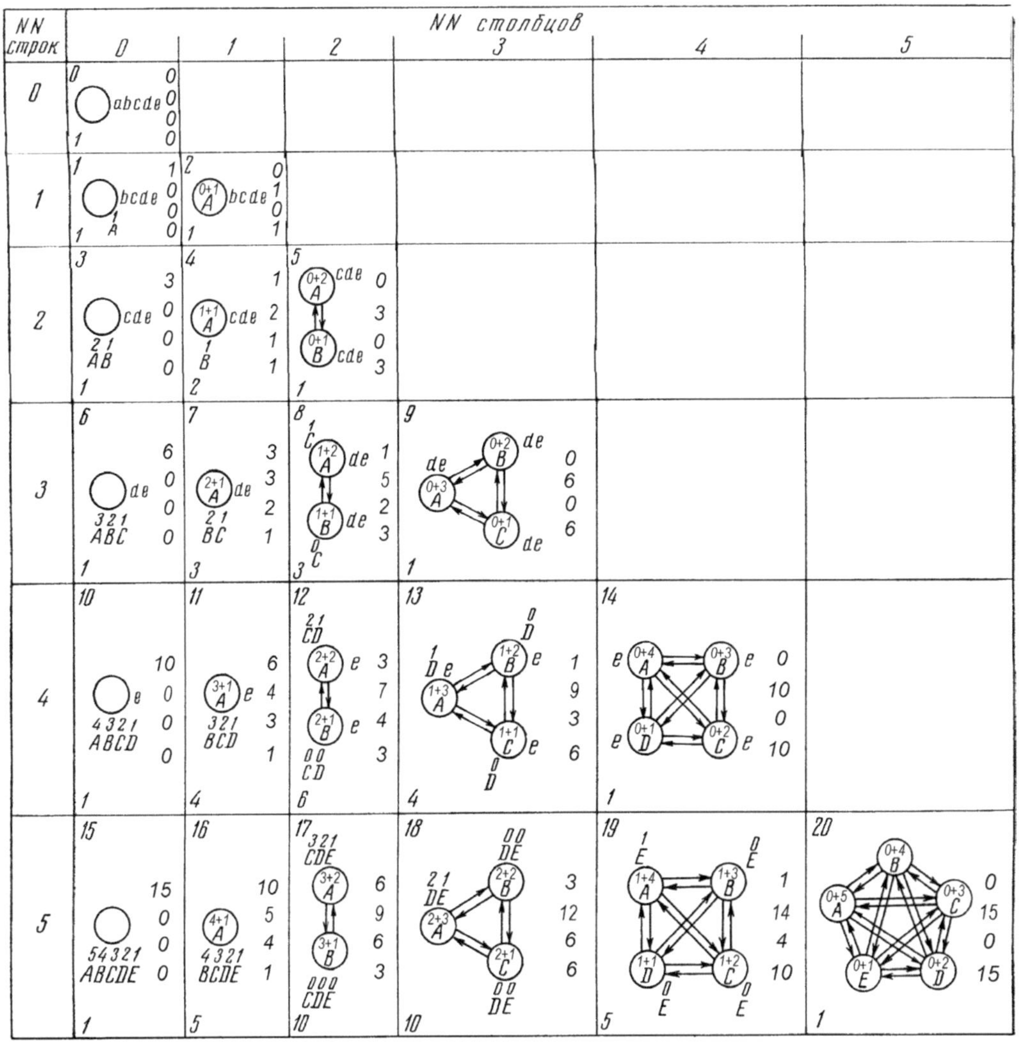

Figure 3. A periodic table of histions with different kinds of the function divisions between cells. Cell 0 of the table contains an independent wide-ranging single-cell organism lacking any potency and performing all functions (denoted by small letters) for himself only. A table row № (m) shows the number of the functions having potencies acquired for specialization. A table column № (n) indicates the number of specialized functions (denoted by capital letters inside a circle) and, correspondingly, of specialized cells. The number of non-realized and realized structural potencies is indicated by left and right numbers cited above letters in circles. The number placed in the upper left-hand corner of the table cell denotes an ordinal number N of a histion. The number in the lower left-hand corner of the table cell is the number H of isotopes varying in the composition of specialized functions but having the same N. The numbers located in the right part of the table cell display from top to bottom the numbers of non-realized and realized generative potencies as well as the numbers of non-realized and realized structural potencies.

The periodic table includes all histions that are possible in the framework of the adopted rules of development. The development itself can be represented as the histion's advance on table columns and rows and along the isotopic coordinate, as well. For example, when moving down the columns, the acquisition of generative potencies occurs, while in moving on the

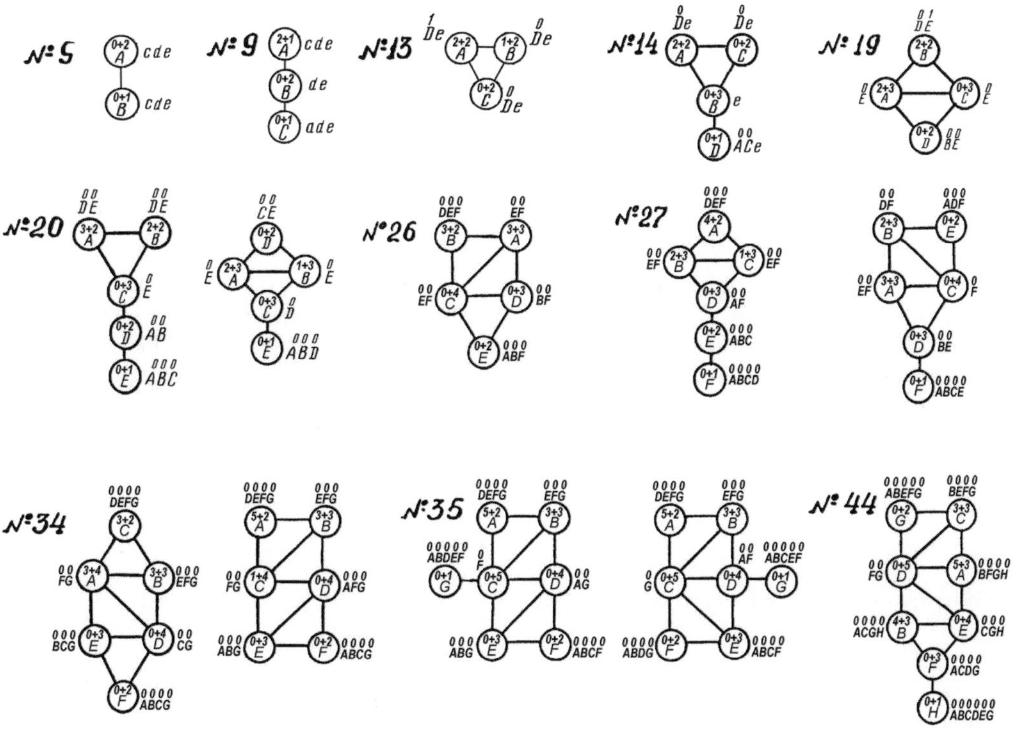

Figure 4. Some examples of coincidence of monomeric histion structures with the spatial organization of cleavage of *C. elegans* in the development. Here, histions are simplified, their numbers agree with those shown in **Figure 4**. Real photographs of the cleavage have been taken from the internet resource http://Wormatlas.org. A—Founder cell; P—Stem cell.

rows to the right, their realization occurs in the progressive development, just representing the histion's macroevolutionary variability. When moving along the isotopic coordinate, a search of isotopes is carried out in the deviant development.

The above-described ideas accord well with the existing knowledge of the basic laws of multicellular organism phylogenesis. Versatile, full of potencies, single-celled ancestors of all subsequent histions are placed in the zero column of the table. When moving along the rows to the right of the table, these potencies are being realized by way of increasing the number of specialized histion cells at the expense of asymmetrical mitosis of the cells' ancestor. Within each line, the depletion of potencies leads first to the isolation of stem cells and then to the completion of the histion development cycle. To ensure the histions' survival, their positions in the middle of line (i.e., within the zone of adaptive maximum) and approaches to it via microevolution shall be the most advantageous ones.

Any continuation of further development as well as the approach to the right end of the line leads to the loss of all specialized members of the histion. Thereby, it causes reverting into the single-cellular state, an additive increase of the number of generative potencies, and the transition to the beginning of the next line. Then everything is repeated again. Thus, the histion development is characterized by such features as: finiteness, directivity, attainment and loss of diversity, parallelism, and repeatability. The same properties are typical for the development of real biological organisms in different taxonomic categories. Therefore, the proposed table serves as a model for building natural systems of real organisms.

As mentioned above, a set of rules may be modified so as to allow the development of histions in variable conditions. For example, when canceling the rule of the equal number of partners for each histion cell, the existence becomes possible of such a set in which the number of cell partners depends on the number of their structural potencies. The reality of such histions is verifiable. Since a direct observation of histions is possible but early in the development (at the cleavage stage, when they exist in the monomeric form), a correlation has been studied between histion models and spatial organization versions in the development of blastomeres (**Figure 4**, such a correlation shows a good agreement of models with the reality).

Other examples were reported earlier [15]. These results confirm the validity of the histion theory. Thus, histions are a new object of developmental biology. They represent an independent and the so-far overlooked level of the biological organization between cells and tissues.

8. The formation of cellular networks

In organism development, a monomeric state of histions is quickly replaced by their polymerized state. This view gives a clue to the elaboration of a predictive theory of biological tissue structures. The basis for this theory is the proposition that it is histions, and not the cells, as such, that are the elementary morphofunctional units of tissues. This understanding is justified by the fact that most of the tissues consist of several, rather than of one, types of cells. We can assume that tissues are the result of polymerizing histions and represent one-, two-, or three-dimensional cellular networks. The regularity of networks can be considered as a manifestation of the histion structure of tissues. Such networks reflect sociology of tissues and characterize tissue properties. Network structures give us information on variants of three-dimensional organization of tissues. However, modern experimental histology does not have any effective methods for studying the three-dimensional structure organization and remains two-dimensional histology. Unfortunately, the existing mathematical theory of the structure of biological tissues [20, 21] fails to put forward the needed set of models. This is why tissue engineering lacks any scientific basis up to the present.

A notion of histions has given insight into a theory of spatial organization of tissues providing a new effective approach to the study of tissue structures [15]. So, based on the known histion composition/structure and using histion polymerization rules, it is possible to calculate a variety of cellular network structures. The cellular networks may be conceived as geometric and topological models of the spatial tissue organization. A set of such models allows one to predict a tissue development and then to experimentally find its histoarchitecture unknown earlier.

So, it has been shown that for a two-dimensional case (single-layer epithelium), the existence of 11 regular models is possible for the cellular networks known as regular parquets of Kepler. We have already found nine of the models in real tissues and predict the detectability of two more [15].

For a three-dimensional case (multilayered tissues), a family of regular models has been constructed too. Using such models in computer simulation by means of special programs, Gistoarkh and Gistored give the chance to animate and visualize cell shapes and their interrelation in the layer space. With the help of the constructed models, any reconstruction of three-dimensional structures of multilayered tissues is reduced to comparison of their cuts with the sections of the models followed by the choice of that from them which well corresponds to reality.

Reconstructing cell shapes and cellular network topologies has been carried out by this method for a number of integumentary and sensory epithelia [15, 22, 23]. At the same time, the reconstruction requires less cuts while the accuracy attainable allows us for the first-time determination of tissue topologies and those variations in development. Regular structures, translational symmetry, and stoichiometric composition are typical for all cellular networks. Also, it has been shown that changes of cellular networks in development are comparable with phase transitions. At the same time, various defects giving some additional properties to tissues are also typical of such networks [24]. By way of illustration, let us compare three tissue models (such as cellular mosaics, a network, and a histion) with the real tissue structure in order to see how the tissue models correlate with it (**Figure 5**).

From this analysis, it can be concluded that further studies into tissue structures should be focused on the elucidation of the composition and structure of their cellular networks and of

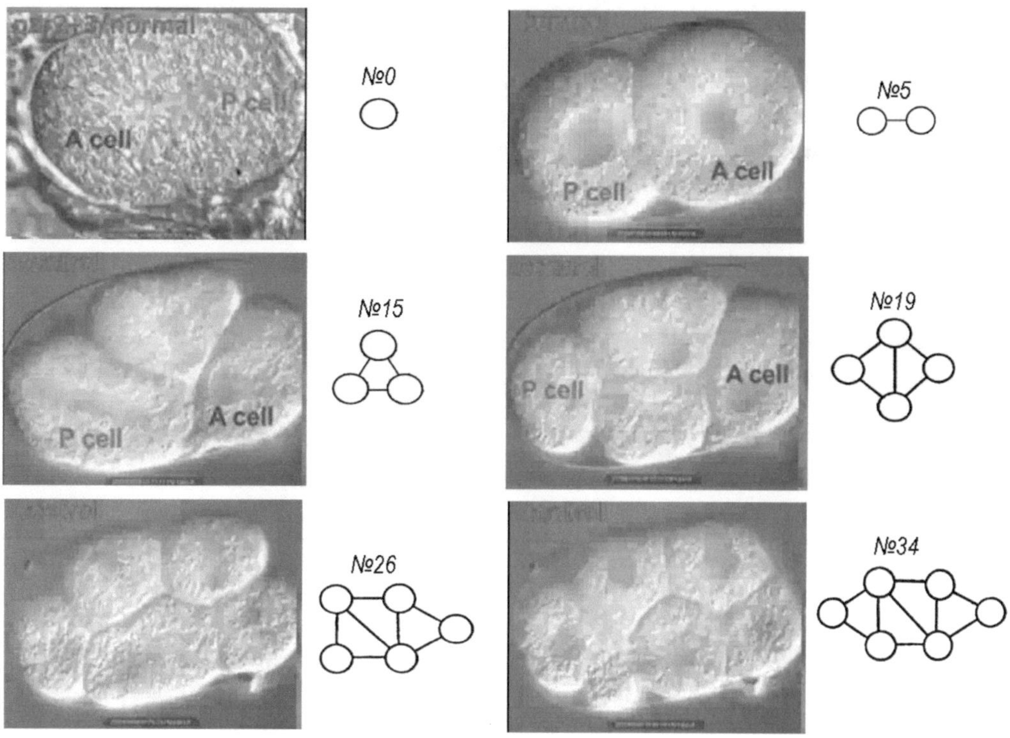

Figure 5. An example of coincidence between the real AB_2 tissue structure (the auditory epithelium of the cochlea of birds) and its model as cellular mosaics, a network, and its histion. Hair cells are dark, supporting cells are light.

histions, as well. The use of such models drastically improves the performance of exploring the spatial organization of the real cell layers. The analysis of networks opens a new line of investigation into the structure of biological tissues. It is of interest to study predictability of their changes both in development and in disease. Moreover, such an approach would become particularly important and promising for tissue engineering. Parameters of cellular networks could serve for these purposes as new diagnostic signs.

9. Economic results of the multicellularity development

The division of labor and the technological progress can lead to ever-increasing productivity and consumption of goods per head. From the times of Adam Smith, just the same result of development was considered normal and preferable in the human community. However, this is not true of the development of multicellular organisms. Despite the fact that the division of functions between cells can, in principle, cause the consumption and production of metabolites per cell to increase, that idea could be implemented in the early stages of development only. So it failed to be widely used later on because a limited number of niches and the biosphere as a whole as well as the unlimited ability of organisms and cells to reproduce themselves ultimately lead to the permanent deficiency of available resources. As it was noted by Malthus, the permanently improved economy of the multicellular organism vitality has become one of the major trends in the organism development. This is one of the fundamental laws of development [25]. In this case, the number of metabolites produced per cell materially decreases. For example, compared with the original single-cell organism, the specific (per cell) production and consumption of metabolites in animals with a body weight of about 100 kg goes down to 10 times and even to 100 times less—in large animals, even more.

And how did the cell need change therewith? To evaluate them, the following should be emphasized: the versatility of the basic molecular mechanisms, such as: cellular bioenergetics, reduplication, transcription, and translation, enables the single-cell organism to spend for synthesizing 1 g of DNA, RNA, or protein as many macroergs of nitrous bases and of amino acids as the higher organism cells spend. Thus, reproductive needs of developing cells remain practically unchanged. In contrast, the needs of cells for performing specialized functions decrease regularly in development. Growing specialization and the accompanying bioengineering progress have made possible the economy's improvement of many processes associated with environmental adaptation, searching for food, and the production of metabolites.

If we compare the metabolites' productivity decline with the constancy of reproductive needs, we shall see that the quantity of metabolites produced in an organism is insufficient to provide reproduction of all cells. This means that the percentage of the cells capable to reproduce themselves is bound to decline. This can be accomplished by a number of measures. Firstly, this is ensured by the very structure of the periodic table because the law of conservation of potencies as well as the depletion of potencies sets a limit to the number of cells involved in the development. Secondly, this is provided by the function order in the development. So a resource-consuming reproduction function is among the first involved in the development. Such a measure causes division of an organism into the reproductive system (constituting the least part by weight) and the soma that is the largest part of the organism. Therewith, the reproductive system is supplied according to a high reproductive norm, whereas the soma

http://dx.doi.org/10.5772/intechopen.73658

switches over to a more economical regime of the specialized vital activity. Thirdly, the soma also generates two different size populations of cells. The smaller one consists of the stem cells and of the fissioning committed cells that are relevant to the beginning of table rows. These populations make up cambia (the local reproductive "incubator" zones) and they are supplied according to the high reproductive norms. The majority of cells are non-reproducible (or working) cells and they are related to the ends of table rows, say much like keratocytes or nucleus-free red blood cells. The supply norms for these kinds of cells are substantially lower. In response to the above-described measures, a share of reproduced cells regularly decreases in the organism development (from 100% at early stages down to less than 1% in the grown-up individual). Obviously, the realization of these measures was accompanied by changing development rules. For example, rule number 1 is no longer applicable to all the cells of the body.

In order to separate various systems and populations according to their blood supply levels, the organism had to work out different physiological compartments in the form of various anatomical organs. Further, to provide uneven blood supply to different compartments, the organism has formed a system for providing both a separated blood supply and a selective uneven distribution of metabolites as the saying is "Some feast, and some fast." The non-uniformity is graphically shown in **Figure 6** using the Lorentz's diagram where the hatched area can be a quantitative measure of divergence between the uniform and the real blood supply to different organs [15].

Therewith, the organism imposes a ban on the free migration of most working cells between compartments. The above-discussed complex of measures enabled the organism to essentially decrease productive consumption expenses per cell and thereby to increase the organism's economical efficiency.

Thus, a non-uniform distribution of metabolites is vitally important for the existence of a developed multicellular organism. From the Lorentz diagram, a trend becomes clear of a spontaneously proceeding process caused by various damaging factors. Such processes are focused on leveling a distribution of metabolites (**Figure 6**, the arrow and the dotted line). That is, the rich

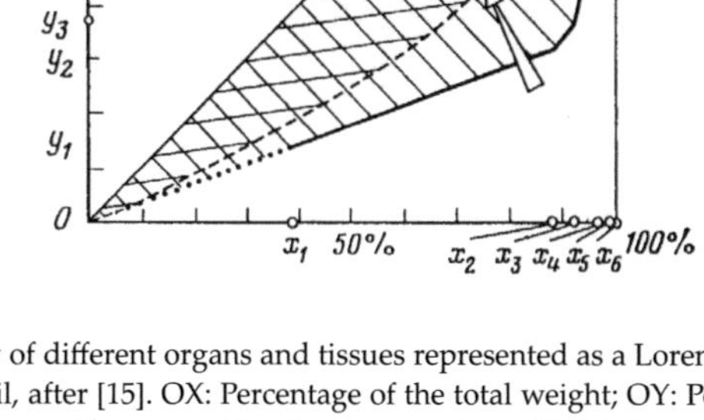

Figure 6. Unequal blood supply of different organs and tissues represented as a Lorentz diagram that is built using the data obtained by Folkov and Nil, after [15]. OX: Percentage of the total weight; OY: Percentage of the total volume per minute. x_1 and y_1—Bone, adipose and/or connective tissues; x_2 and y_2—Skeletal muscles; x_3 and y_3—Skin; x_4 and y_4—Digestive apparatus; x_5 and y_5—Brain; and x_6 and y_6—Heart. The arrow indicates a development trend of spontaneous processes which decrease inequality of distribution.

will grow poorer, while the poor will grow slightly rich. The redistribution of metabolites can be revealed in various kinds of pathologies, such as atrophies and/or inflammations. But if such redistribution is carried out in combination with the redistribution of potencies, it can reveal itself as an uncontrollable increase of some kinds of cells followed by their arbitrary settling down in different compartments (organs). This effect takes place in the malignant tumor growth with metastasis. Full equalization of metabolites distribution corresponds to the death of the organism.

So, to be able to survive at scarce resources, multicellular organisms had to take a number of measures, such as:

- the caste division of cell populations into those having potencies for developing and those lacking them, when the reproduction is not allowed for most cells;
- non-uniform distribution of metabolites;
- forming of special "incubators" for carrying out a controllable centralized production of just as many specialized cells as it is required to conform to the vacancies available;
- regulated choice of a profession; and
- a strict ban on the free migration of most working cells.

And these measures have made it possible to get the overall result of the progressive development including technological progress and improving of the vital activity economy.

10. Conclusion

Division of functions between cells generates a multicellular organism with its inherent sociology and economy, wherein there is no room for the equality of cells. These developmental aspects have their own mechanisms and are regulated according to their own laws, unknown so far. They should become the subject of a special study. Within the framework of the study, it is necessary to tackle such problems as:

1. determination of L, m, and n parameters which are inherent in multicellular organisms;
2. elucidation of mechanisms and of the sequence of obtaining both generative and structural potencies and of their realization, as well as the determination of the number of the potencies and of those dynamic changes in development;
3. investigation into monomeric histion compositions/structures and construction of their parametric systems;
4. investigation of cellular networks and elaboration of three-dimensional histology;
5. quantitative analysis of metabolites' distribution in a real organism.

The resolution of these issues will become a step forward to the development of a nomogenetic theory that would allow not only modeling but also measuring of the multicellularity development (both in norm and in disease). Also, this would enable development to be controllable thus providing the scientific basis for tissue engineering and regenerative medicine.

Acknowledgements

This work was supported by the state assignment of FASO of Russia ("Mechanisms of formation of physiological functions in phylo- and ontogenesis as affected by endogenous and exogenous factors").

Author details

Gennady A. Savostyanov

Address all correspondence to: genasav38@mail.ru

Sechenov Institute of Evolutionary Physiology and Biochemistry of the Russian Academy of Sciences, St. Petersburg, Russian Federation

References

[1] Chandebois R. Cell sociology: A way of reconsidering the current concepts of morphogenesis. Acta Biotheoretica. 1976;**25**(2):71-102

[2] Gass G, Hall BK. Collectivity in context: Modularity, cell sociology, and the neural crest. Biological Theory. 2007;**2**(4):349-359

[3] Robinson CV, Sali A, Baumeister W. The molecular sociology of the cell. Nature. 2007;**450**(7172):973-982. DOI: 10.1038/nature06523

[4] Fredericks M, St M, Odiet JA, Fredericks JP. Towards an understanding of cellular sociology and its relationships to cellular biology. Education. 2003;**124**(2):237-256

[5] Nowak MA. Five rules for the evolution of cooperation. Science. 2006;**314**:2560-2563. DOI: 10.1126/science.1133755

[6] Grosberg RK, Strathmann RR. The evolution of multicellularity: A minor major transition? Annual Review of Ecology, Evolution and Systematics. 2007;**38**:621-654. DOI: 10.1146/annurev.ecolsys.36.102403.114735

[7] Aktipis A. Principles of cooperation across systems: From human sharing to multicellularity and cancer. Evolutionary Applications. 2015;**9**(1):17-36. DOI: 10.1111/eva.12303

[8] Pavlogiannis A, Chatterjee K, Adlam B, Nowak MA. Cellular cooperation with shift updating and repulsion. Scientific Reports. 2015;**5**(17147):1-12. DOI: 10.1038/srep17147

[9] Cox RP, Krauss MR, Balis ME, Dancis J. Metabolic cooperation in cell culture: Studies of the mechanisms of cell interaction. Journal of Cellular Physiology. 1974;**84**:237-252. DOI: 10.1002/jcp.1040840210

[10] Pulendran B. Division of labor and cooperation between dendritic cells. Nature Immunology. 2006;**7**(7):699-700. DOI: 10.1038/ni0706-699

[11] Welner R, Amabile G, Bararia D, et al. Normal stem and progenitor cell sociology within the leukemic microenvironment (HEM3P.300). The Journal of Immunology. 2014;**192** (1 Suppl):51

[12] Zhang MX, Chen JJ, Wang L, et al. Preliminary analysis of cellular sociology of co-cultured glioma initiating cells and macrophages in vitro. Translational Neuroscience and Clinics. 2016;**2**(2):77-86. DOI: 10.18679/CN11-6030/R.2016.018

[13] Wilson EO. Sociobiology. Cambridge, MA: Harvard Univ. Press; 1975. 697 p

[14] Requejo RJ, Camacho J. Evolution of cooperation mediated by limiting resources: Connecting resource based models and evolutionary game theory. Journal of Theoretical Biology. 2011;**272**:35-41. DOI: 10.1016/j.jtbi.2010.12.005

[15] Savost'yanov GA. Essentials of Structural Histology. The Spatial Organization of Epithelia (in Russian). Sankt Petersburg: Nauka; 2005. 375 p

[16] Savost'yanov GA. Unified approach to estimation of progressive and deviant development based on the analysis of division of labor. Paleontological Journal. 2015;**49**(14):1667-1680. DOI: 10.1134/S0031030125140254

[17] Savostianov GA. Emergence of stem cells. The development of multicellularity and its quantitative characteristics (in Russian). Cytologia. 2016;**58**(8):577-593

[18] Lyubishchev AA. On the form of a natural system of organisms. In: Problems of Form, Systematics and Evolution of Organisms (in Russian). Moscow: Nauka; 1982. pp. 24-36

[19] Popov IY. Periodic System and the Periodic Law in Biology (in Russian). Comp. Sci. Pres. KMC: Moscow; 2008. 223 p

[20] Smolyaninov VV. Mathematical Models of Biological Tissues (in Russian). Moscow: Nauka; 1980. 368 p

[21] Maresin VM. Spatial Organization of Embryogenesis (in Russian). Moscow: Nauka; 1990. 169 p

[22] Magnitskaya YG, Grefner NM, Golubeva TB, Vorobyov AV, Levchenko VF, Savostyanov GA. Epithelium 3D structure transformation during development, with bird basilar papilla sensory epithelium as an example (in Russian). Sensory System. 2009;**23**(4):334-345

[23] Savost'yanov GA. The origin of elementary units of multicellularity and development of a spatial organization of cell layers. Biology Bulletin. 2012;**39**(2):127-136. DOI: 10.1134/S1062359012020100

[24] Savost'yanov GA. Structural defects in regular cell mosaics and their putative functional significance. Paleontological Journal. 2014;**48**(11):1252-1162. DOI: 10.1134/S0031030114110100

[25] Zotin AI. Thermodynamic Approach to Development, Growth and Aging (in Russian). Moscow: Nauka; 1974. 184 p

Chapter 11

Protein Reabsorption in the Amphibian Kidney: Comparative and Evolutionary Aspects

Elena V. Seliverstova and Natalya P. Prutskova

Additional information is available at the end of the chapter

http://dx.doi.org/10.5772/intechopen.73659

Abstract

Protein reabsorption in the renal proximal tubule (PT) is a vitally important process which prevents the loss of filtered proteins and provides their participation in subsequent metabolism. Despite considerable changes in renal structure and function in the process of evolution, very little is known about the functional similarities or specifics of tubular protein reabsorption in the kidney of lower vertebrates compared with the mammalian and human kidney. This article presents an overview of our recent studies on protein reabsorption in the kidney of amphibians, which are used as one of the main animal models for current biological and biomedical research. In frogs, newts, and rats, the absorption capacity of epithelial PT cells was studied after the introduction of green fluorescent protein (GFP), yellow fluorescent protein (YFP), and lysozyme. Molecular mechanisms of receptor-mediated protein endocytosis were also investigated by immunohisto- and immunocytochemistry, electron, fluorescent, and laser scanning confocal microscopy.

Keywords: amphibians, cubilin, comparative physiology, endocytosis, evolution, frog, kidney, megalin, protein reabsorption, proximal tubule

1. Introduction

Renal protein reabsorption is a process which reduces urine protein excretion and allows the absorbed proteins to participate in subsequent metabolism. It also provides the retrieval of other specific substances including the conservation of carrier-bound vitamins. Detailed investigations of this process are of great importance for understanding renal physiology, tubular disorders, and homeostatic control mechanisms. Reabsorption of filtered proteins occurs in the epithelium of proximal tubule (PT). Despite the high absorption capacity of mammalian PTs, increasing protein uptake and prolonged overload situation may lead to tubular proteinuria and tubule injury and subsequently may induce tubular interstitial damage [1, 2]. Obviously, extensive studies of tubular protein reabsorption are of great importance for renal physiology and pathology of proteinuric diseases in humans and mammals. Probably for this reason, such studies are limited by primarily clinical investigations and use of theoretical and mammalian animal models, although a novel model was also proposed using the axolotl [3]. To date, the process and mechanisms of protein reabsorption in mammalian PTs have been rather well explored at the cellular and molecular level. At the same time, very little is known about tubular reabsorption and endocytic transport of proteins in the non-mammalian kidney. Despite considerable and progressive transformation of vertebrate renal structure and function in the process of evolution (e.g., see [4]), the structure and function of PTs do not appear to have undergone major evolutionary changes [5]. It can be assumed that PT functions are mostly evolutionarily conserved, but this assumption does not have sufficient experimental foundations, at least in relation to protein reabsorption. There is no sufficient information about the degree of the similarities or differences in tubular protein uptake, molecular mechanisms, and regulation of this process in the ascending series of the vertebrates or during ontogeny. Our interest in the study of renal protein reabsorption in the amphibian kidney is due to several reasons. Amphibians occupy a key position in the evolution of terrestrial vertebrates and bridge the gap between the aquatic fishes and the terrestrial vertebrates. The basic renal physiology of these poikilothermic tetrapods is relatively well understood, primarily with regard to water and ion transport but not to tubular protein uptake. Along with that, amphibians as animal models are one of the main objects of current biological and biomedical research. In this chapter, we present a brief survey of available information about tubular protein reabsorption and molecular mechanisms of protein endocytosis in the kidney of amphibians, based on our research within the context of existing literature and current ideas about molecular and cellular mechanisms of endocytosis. Some comparative and evolutionary aspects of the issues involved are also considered.

2. Structural and functional basis of glomerular filtration and tubular protein reabsorption in the amphibians

Vertebrate kidneys develop via three successive stages of formation in the process of evolution or during ontogeny—pronephros, mesonephros, and metanephros. Pronephros constitutes the mature kidney in most primitive vertebrates (cyclostomes); it is the earliest stage in fishes and tetrapods and the functional embryonic kidney in amphibians. Mesonephros is the permanent kidney of amphibians and most fish, replacing the pronephros of the embryonic and larval stages. It serves as the main excretory organ of aquatic vertebrates and as a temporary kidney in reptiles, birds, and mammals. During embryogenesis in amniotes, pronephros is succeeded by the mesonephros, which gradually degenerates, and a more complex metanephros arises caudal to the mesonephros and develops as functional adult kidney of higher vertebrates. The nephron is the basic structural and functional unit of the kidney. Glomeruli, proximal, and distal segments as major parts of the kidney nephrons are present in nearly all vertebrates. It is known that the filtration properties of the glomeruli are determined by the pore size of the filtration barrier and depend on the physical–chemical properties of plasma proteins. The cut-off molecular mass for filtration of plasma proteins in renal glomeruli during normal conditions has generally been assumed to be lower than the molecular mass of serum albumin and some other large proteins (in the range of 60–85 kDa). Structure of the filtration barriers within the glomeruli of studied amphibians and mammals is very similar [6, 7]. It concerns the ultrastructure of the glomerular wall, in particular capillary endothelium, basement membrane, endothelial cell layer, and slit diaphragm, limiting permeability.

According to our morphological studies, the glomerular filtration barrier in the kidney of *Rana temporaria* showed the classic three-component structure (**Figure 1A**), as described in other anuran and urodel species [8–10]. It is composed of a layer of capillary endothelial cells facing the blood, a heterogeneous glomerular basement membrane, and a visceral epithelial cell layer which faces the urinary space of Bowman's capsule. Mesangial cells are distributed between the capillary loops. The ultrastructure of the PT has also typical features of these parts of the nephron [8, 9, 11, 12]. Well-preserved endocytic apparatus, including vesicles, dense apical tubules, endosomes, and lysosomes (**Figure 1B**), indicates the active uptake capacity of the PT cells [8, 13].

Studies of proximo-distal patterning of the nephrons in the frog, *Xenopus laevis*, showed the presence of early physiological specialization of PTs at the stage of pronephros [14]. In particular, it was found that pronephric PT has an early and a late segment and different transporters are expressed within unique subdomains similar to those in mammalian metanephric PT. The ability of tadpole pronephros to filter and reabsorb fluorescently tagged proteins (serum albumin, codfish parvalbumin) was also revealed after cardiac injections.

In our morphophysiological studies, we investigated the mechanisms of protein reabsorption in the amphibian mesonephros by the methods of fluorescent and confocal microscopy, immunohisto-, and immunocytochemistry. Experiments were performed on common frogs (*Rana temporaria*), newts (*Triturus vulgaris*), and also Wistar rats (*Rattus norvegicus*) for some comparisons.

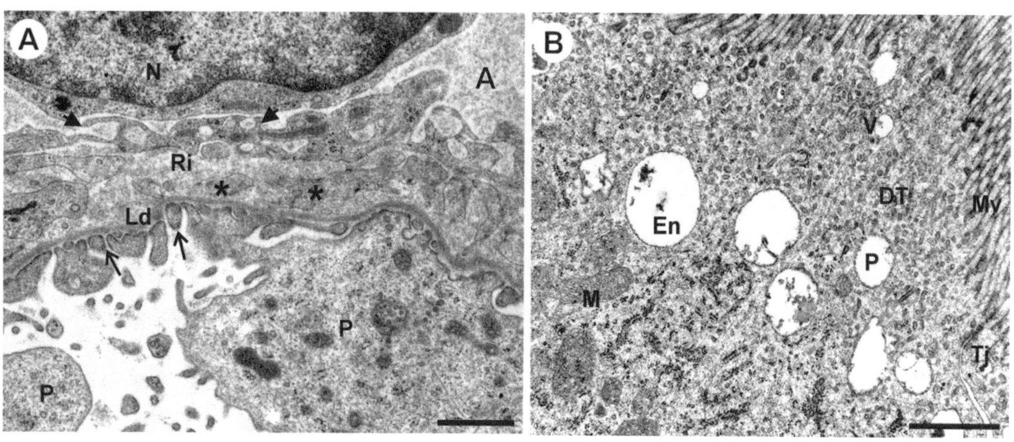

Figure 1. The ultrastructure of glomerular filtration barrier (A) and apical area of proximal tubular cell in the kidney (B) of the frog *Rana temporaria*. Arrows point to the foot processes of podocytes covering thin lamina rara externa; asterisks show the processes of mesangial cells, arrowheads demonstrate numerous fenestrae of the endothelial cells. En, endosome; DT, dense apical tubules; Ld, lamina densa; M, mitochondria; Mv, microvilli; N, nucleus; P, podocyte; Tj, tight junction; V, vesicle. Scale bar 1 μm. Author's drawings.

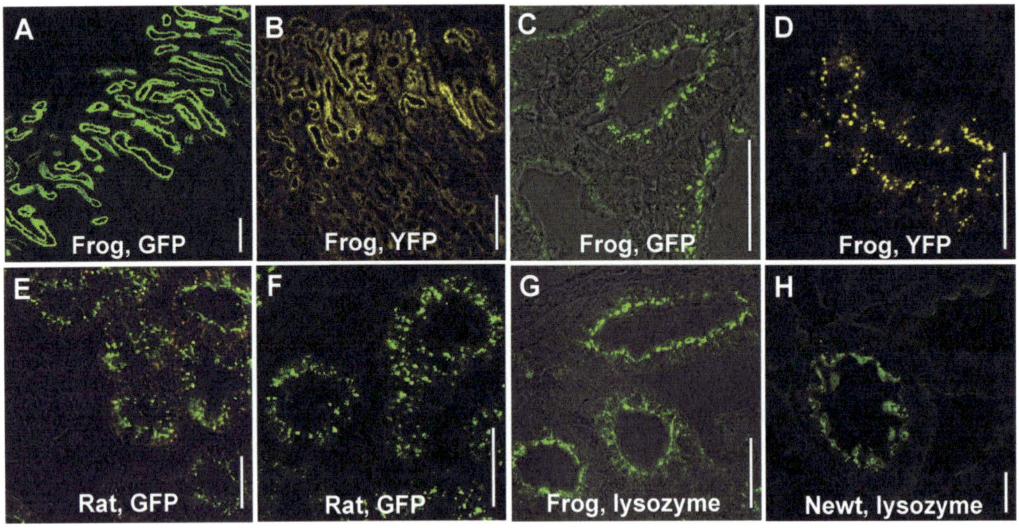

Figure 2. Protein reabsorption in the proximal tubules of the amphibian and rat kidney. In panels: The uptake pattern of GFP, YFP, and lysozyme; 30 min (A–D, F, G), 5 min (E), and 20 min (H) after protein introduction. Scale bars: 100 μm (A, B), 25 μm (C–H). Author's drawings.

We used recombinant fluorescent proteins, green fluorescent protein (GFP), and yellow fluorescent protein (YFP), which turned out to be freely filtered in glomeruli and reabsorbed in epithelial cells of PTs after intravenous injections in frogs, as in rats [15]. Earlier, intestinal absorption and the resulting accumulation of these fluorescent proteins in renal PTs after their intragastric administration were established [16]. Later, the uptake of lysozyme in the amphibian kidney was also demonstrated [17]. Some examples of tubular reabsorption of abovementioned proteins in amphibians and rats are shown below (**Figure 2**).

3. Protein uptake pattern in the amphibian and rat kidney and quantification of protein reabsorption

In our studies, protein uptake was analyzed after intravenous (i.v.) protein introduction in immobilized (double-pithed) frogs and anesthetized rats. Intraperitoneal (i.p.) injection and subcutaneous (s.c.) introduction (into dorsal lymph sac) were applied to mobile amphibians. Before and during experiments, amphibians were in terms of optimal hydration. Absorbed GFP or YFP was detected in fixed kidney slices by fluorescent or laser scanning confocal microscopy. In frogs, these proteins were revealed in epithelial layer of PT profiles situated in the dorsolateral part of the kidney including supraglomerular zone and superficial areas (**Figure 2A** and **B**). Initially diffuse, a specific signal was visualized in endocytic vesicles of PT cells 10–30 min after protein injections. Bright fluorescent vesicles were located predominantly in apical cytoplasm near brush border and also in perinuclear areas (**Figure 2C** and **D**). In rats, the fluorescent PT profiles were revealed in periglomerular areas of the rat kidney cortex

(**Figure 2E**); the distribution of the protein-containing vesicles was similar to that seen in frogs (**Figure 2F**). Renal uptake of lysozyme in frogs and newts (after i.v. and i.p. injection, accordingly) was proved by immunohistochemistry using rabbit anti-hen lysozyme (as primary antibody) and Alexa Fluor 488 conjugate (secondary goat anti-rabbit IgG conjugated with Alexa Fluor 488). Intracellular distribution of labeled lysozyme in a vesicular compartment of PT cells does not differ fundamentally from the uptake pattern of GFP and YFP (**Figure 2G** and **H**). To discover regularities in protein reabsorption we used a variety of approaches for quantification of protein uptake [15]. After GFP introduction at the doses 0.034–34 μg/100 g body weight, reabsorption of this protein in the kidney was dose-dependent in both frogs and rats [15]. The specific fluorescence intensity, maximum fluorescence, and fluorescence density increased in response to increasing doses of GFP and a high positive correlation was revealed. Reabsorption of fluorescent proteins was also time dependent [15, 18]. With increasing time after injection, there was an accumulation of vesicles with GFP or YFP and a movement of some fluorescent endocytic vesicles from the apical cytoplasm to perinuclear and basal areas. As shown in our recent studies, the number of the formed fluorescent endocytic vesicles is the most suitable and rather adequate parameter for quantitative morphological analysis of the protein absorption rate over a fixed period of time [19]. The dynamics of the accumulation of various proteins in renal PT cells within 30 min after i.v. injections were generally similar (**Figure 3A**) and prolonged for fluorescent proteins in frogs (**Figure 3B**).

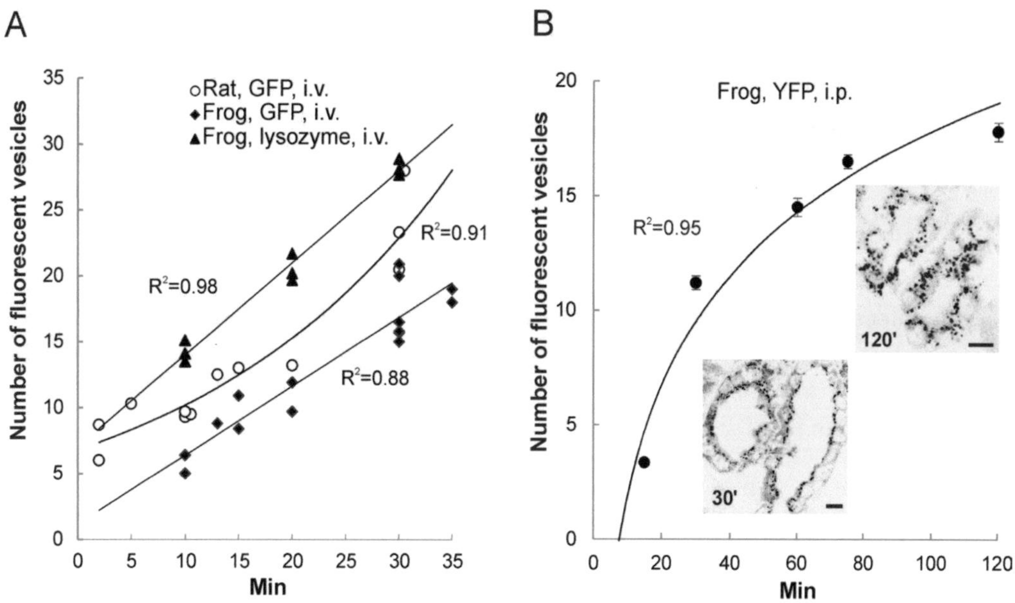

Figure 3. Time-dependent protein uptake pattern in the frog and rat kidney. Ordinate: The average number of protein-contained vesicles (per five neighboring epithelial cells). (A) The uptake of GFP and lysozyme in rats and frogs; (B) the absorption dynamics of YFP in frogs; R^2, approximation confidence. Inverted confocal images of tubule profiles show the accumulation fluorescent endocytic vesicles with increasing time after YFP injection. Scale bars: 10 μm. Author's drawings.

4. Comparative analysis of renal protein uptake in the rat and frog kidneys

Known stability of the molecular structure of GFP and YFP and their resistance to lysosomal degradation give an advantage in detection of completely absorbed proteins, allowing us to estimate the uptake and intracellular transport of intact protein molecules for quite a long time. Using these proteins, we focused our research on functional differences in the kidneys of mammals and amphibians based on more detailed comparative analysis of protein uptake. Despite the similarity of the basic patterns of tubular protein absorption in rats and hydrated frogs after injections of equal doses of different proteins, it is obvious that tubular handling of GFP is quicker in rats compared to frogs (**Figure 3A**). Granular fluorescence in rat PT cells appeared 2–5 min after GFP introduction (**Figure 2E**). In 30 min, the number of vesicles with internalized protein significantly increased and fluorescent vesicles scattered over the epithelial cell cytoplasm (**Figure 2F**). However, in the subsequent period, the number of GFP-containing vesicles in PT cells and the means of maximum fluorescence have dramatically decreased [15, 20]. In 1–2 h, green fluorescent vesicles were absent in the vast majority of PTs. Since the disappearance of GFP signal signifies the destruction of its molecular structure, it is reasonable to suggest that in the rat kidney GFP is metabolized in some fashion and at least a partial lysosomal degradation of this protein occurs.

In contrast to rats, in the frog PT cells intracellular transport GFP and YFP and the gradual accumulation of fluorescent vesicles took place for a long time and had a similar character, regardless of the way of protein introduction [15, 20, 21]. It means that in frogs these proteins are filtered at a relatively slow rate and remain in circulation for quite a long time. Absorbed fluorescent proteins migrated from apical cytoplasm to perinuclear zone only in 40–60 min. Process of absorption and accumulation of injected protein lasted for 1.5 h and then ended (**Figure 3B**), without reduction in the number of fluorescent vesicles as a sign of protein degradation.

As shown in our comparative physiological study of renal functions, in hydrated frogs an intense water diuresis occurs, in contrast to rats whose values of osmotic free water clearance indicate antidiuresis [20]. Despite the active fluid filtration, the protein absorption rate in the frog kidney was substantially slower than that in the rat kidney. This can be due to a slower glomerular filtration rate (GFR), resulting in a longer period of protein circulation in blood and in prolonged tubular protein reabsorption. According to our research, creatinine clearance-measured GRF in frogs is 0.028 ml/min, that is about 8 times slower than GRF in rats. Generally, GFR in the kidney of *Rana temporaria* corresponds to the range of this parameter measured for a number of tailless and tailed amphibian species. Specifically, it most closely approximates the GFR values in such amphibian species as the Chilean toad (*Calyptocephalella gayi*), clawed frog (*Xenopus laevis*), and northern leopard frog (Rana pipiens)—0.031, 0.05, and 0.056 ml/min, respectively (see [22]). Thus, the peculiarities of the protein uptake revealed in the frog kidney, as compared with the rat kidney, consisted of a lower protein reabsorption rate, intracellular distribution of internalized protein in cytoplasmic compartments, and protein degradation rate.

5. Hormonal modulation of protein uptake in the frog kidney: effect of arginine vasotocin (AVT)

The contribution of the glomerular activity to the process of renal protein reabsorption may be very important for amphibians because in their mesonephros, unlike the mammalian metanephros, the degree of diuresis highly depends on the blood flow through the glomerular capillaries. The role of glomerular filtration in controlling the volume of extracellular fluid differs markedly in lower and higher vertebrates [23–25]. For instance, in fish, amphibians, and reptiles, GFR is not constant, and diuresis depends on variable or intermittent glomerular filtration, in contrast to birds and mammals [24]. In fish and poikilothermic tetrapods, tubular water reabsorption is far less variable compared with that of higher vertebrates [23]. In semiaquatic frogs, urine flow is the greatest in hydrated animals and reduced during dehydration [23, 24], and GFR is hormone dependent [25]. Arginine vasotocin (AVT) causes a reduction of GFR by constricting the preglomerular arteries [26–28]. The glomerular action of AVT is supported by the location of vasotocin receptors subtype 1 over the glomeruli in the amphibian kidney [27, 29].

We suppose that AVT-induced decline in GRF and following reduction of tubular fluid flow can hinder the transfer of proteins to their binding sites on the luminal membrane of frog PT cells and reduce the rate of protein reabsorption. To study the effect of AVT on tubular protein reabsorption in hydrated frogs, we estimated the pattern of GFP uptake after preliminary injections of this hormone [15, 30]. When AVT (0.1 fmol–1 nmol) was introduced 20 min before GFP, reabsorption of injected protein decreased in a dose-dependent manner. At the dose over 1 pmol, AVT provoked irregular GFP uptake pattern and the clusters of fluorescent PTs were observed in only some dorsolateral parts of the kidney. Absence of differences in GFP reabsorption between frogs after injections of low AVT doses and control animals suggests that in hydrated frogs, at room temperature and without osmotic stimulus, most of the glomeruli are continually active. Uneven distribution of fluorescent PT profiles may be a consequence of a decrease or complete cessation of filtration in individual glomeruli. The data suggests that not all of the glomeruli or preglomerular vessels are equally responsive to AVT. To insure whether AVT-induced reduction of GFP uptake is a consequence of the hormone effect on the vascular tone, a V1a receptor antagonist was applied [15, 30]. Administration of V1 antagonist (0.01–1 nmol) 10 min before AVT significantly increased GFP uptake reduced by the action of AVT. Thus, in *Rana temporaria*, AVT may indirectly modulate the tubular protein transport and its effect is mediated by V1a-like receptors.

6. Molecular and cellular mechanisms of endocytosis in the amphibian kidney

In mammalian and human kidneys, the filtered proteins are reabsorbed in PT cells by receptor-mediated endocytosis, then are transferred into endosomes, and finally to lysosomes for degradation. According to modern concepts, this process involves two main membrane receptors, megalin (megalin/lrp2) and cubilin, and also amnionless, and their coordinated action-mediated internalization of different proteins [1, 31]. Existence of genes for megalin, cubilin, and amnionless in Xenopus genome was established and the expression of these receptors in Xenopus tadpole pronephros was discovered [32], suggesting their participation in endocytic protein uptake in amphibians. In other lower vertebrates, megalin- and cubilin-dependent endocytosis was shown for the zebrafish pronephros [33].

In our studies, the expression of endocytic receptors in PT cells of the frog mesonephros was revealed after injections of YFP [18] and lysozyme [17] using polyclonal antibodies against megalin and cubilin. In 15–30 min, absorbed YFP was colocalized with immunolabeled megalin or cubilin in apical endocytic vesicles (**Figure 4A–D**).

In the process of time-dependent lysozyme absorption during 10–30 min, similar internalization of megalin, cubilin, and lysozyme was revealed in frogs and also in newts. After protein injections, receptor-specific signals were initially distributed diffusely, along the base of the brush border (**Figure 4E**), and then became more intensive and punctate, in the subapical area of PT cells (**Figure 4F**). So, the involving endocytic receptors in the tubular uptake and vesicular protein transport in the amphibian kidney were proved. No detectable receptor signal was found in PTs of control animals. This indicated to a ligand-induced process of endocytosis with participation of megalin and cubilin, as also noted for zebrafishes [33]. In order to identify the early step of lysozyme internalization, antibody against clathrin was used. This adaptor protein was detected in most PT profiles (**Figure 4G**). The availability of clathrin in most of the PTs of both control and lysozyme-injected frogs confirms the data about constitutive expression of clathrin and its involvement in the continuous uptake of essential nutrients in mammalian cells [34]. Initial colocalization of clathrin and lysozyme and following divergence of both signals were detected (**Figure 4H** and **I**), pointing the movement of lysozyme to the early and

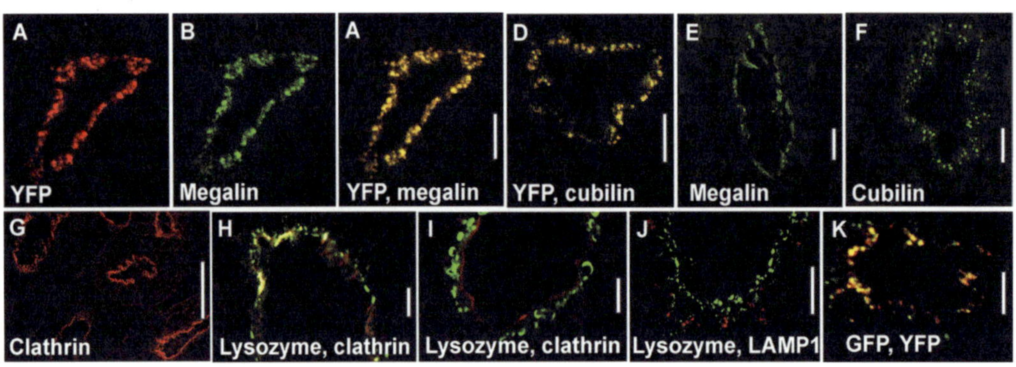

Figure 4. Demonstration of YFP and lysozyme internalization rote in the frog proximal tubular cells by immunohistochemistry and confocal microscopy. In panels: absorbed YFP (A), immunolabeled megalin (B), colocalization of their red and green signals (yellow) on merged image (C), and the same for YFP and cubilin (D); immunodetection of megalin (E), cubilin (F), and clathrin (G) after lysozyme introduction; colocalization (yellow) of lysozyme and clathrin (H) and following divergence of their green and red signals (I), merged; double-labeling of lysozyme (green) and LAMP1 (red), merged (J); the combined uptake of YFP (red) and GFP (green) and their colocalization (yellow) after injection of GFP 1 h before YFP (K), merged. Scale bars: 10 (A–F, H–K) and 40 μm (G). Author's drawings.

late endocytic compartment within 20–30 min. When lysozyme and lysosomal marker LAMP1 antibodies were used, there was no convergence of immune signals (**Figure 4J**). So, lysozyme was retained within endosomal compartment during this time, in contrast to the faster protein traffic in the mammalian kidney. It may be connected with lower metabolic rate and following inhibition of the intracellular transport in hibernating frogs [17].

The results of our immunohistochemical studies of the mechanisms of protein endocytosis were confirmed by immunocytochemistry. Immunoelectron microscopy revealed more detailed intracellular localization of GFP, lysozyme, endocytic receptors, and clathrin 10–30 min after protein injections (**Figure 5**).

As shown by the distribution of gold particles, absorbed proteins can be detected in the apical cytoplasm underneath the brush including intermicrovillar space, in small apical vesicles and large endosomes (**Figure 5A** and **B**). Similar label distribution including intermicrovillar invaginations of luminal membrane and vesicular structures was typical for cubilin, megalin, and clathrin (**Figure 5C–E**). Immunodetection of clathrin proved the internalization of lysozyme via clathrin-coated vesicles. Overall, we provided the evidence that protein reabsorption in the frog mesonephros occurs by receptor-mediated clathrin-dependent endocytosis.

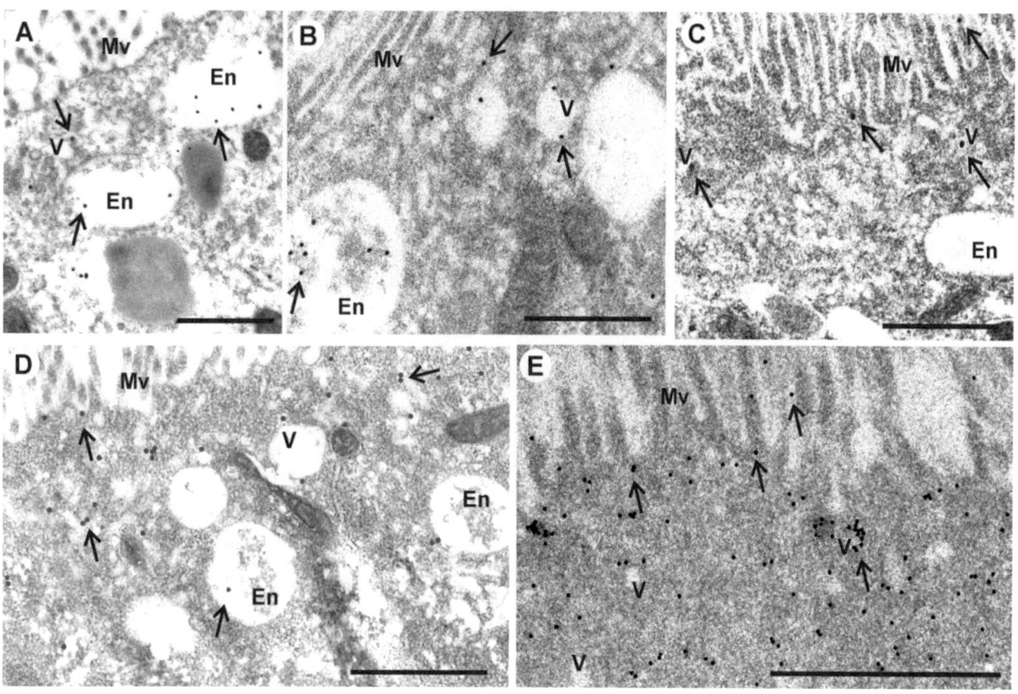

Figure 5. Electron micrographs of the apical region of proximal tubule cells from protein-injected frogs. In panels: immunostaining of GFP (A), lysozyme (B), cubilin (C), megalin (D), and clathrin (E). En, endosome; Mv, microvilli; V, vesicle. Arrows shows the distribution of gold particles. Scale bar: 1 μm. Author's drawings.

7. Effects of combined protein injections and previous protein loading

According to existing data, the results of in vivo and in vitro studies concerning the selectivity and competition of tubular reabsorption of proteins in mammals do not always have a clear explanation and are not well understood. As YFP and GFP are filtered and absorbed in the kidney in the same way, these proteins may be competitive in the absorptive process. We examined the uptake and intracellular traffic of both GFP and YFP under different conditions for competitive absorption in vivo after simultaneous and sequential introduction of equal amounts of these proteins [18, 19]. After simultaneous introduction of GFP and YFP, predominantly colocalized fluorescent signals indicated accumulation of both proteins in the same endocytic vesicles. When two proteins were injected in sequence, one before the other or vice versa, the second protein can be colocalized with the first protein but also located in individual endosomes (**Figure 4K**) because most of the vesicles containing the first protein moved from the apical cytoplasm to other cell areas. Effect of combined injections did not depend on the order of GFP and YFP introduction [18]. Therefore, the total result is shown below (**Figure 6A**). With increasing time interval between injections, a progressive accumulation of the first protein was viewed in 60 and 120 min compared with control (30 min after injection of this protein alone). So, the second protein should be more competitive in the process of the

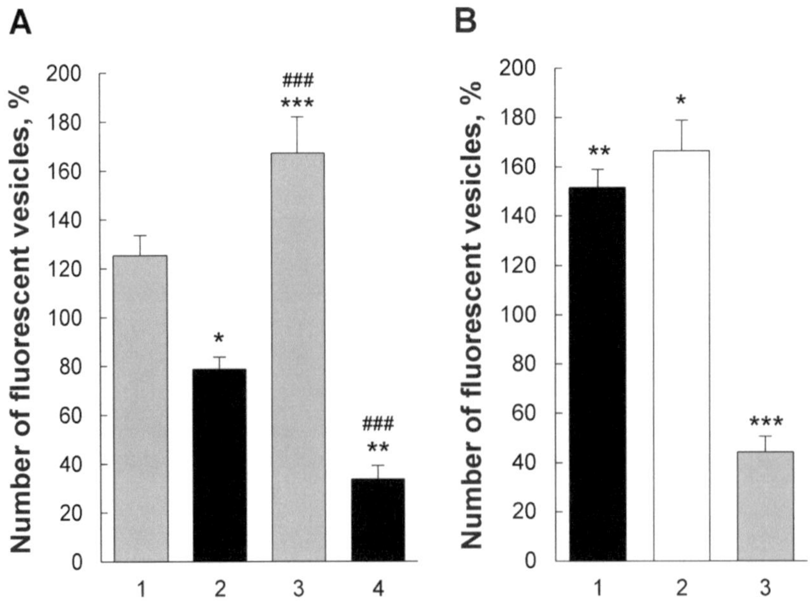

Figure 6. Changes in absorption capacity of frog renal proximal tubule epithelial cells. (A) The effect of combined introduction of two fluorescent proteins: pre-injected protein (1, 3) and the other protein (2, 4) introduced after 30 and 90 min, respectively; (B) the result of immunofluorescence detection of megalin (1), cubilin (2), and lysozyme (3) on the fifth day after cessation of lysozyme loading. Significant differences: $^{*}p < 0.05$, $^{**}p < 0.01$, $^{***}p < 0.001$, compared with control (A) or the third day after stopping the load (B); $^{\#\#\#}p < 0.001$, compared with 1 and 2, respectively (one-way ANOVA followed by a Newman–Keuls test). Author's drawings.

absorption because significant amount of the first protein have been already absorbed in PT cells. However, the second protein uptake decreased.

The results demonstrate the availability of the mechanism capable to limit in vivo the absorption capacity of renal PT cells in frogs. The physiological implication of this downregulation is unknown. In mammals, it may be due to deficiency of endocytic receptors on the apical plasma membrane of PT cells and linked to changes in the initial steps of endocytosis, as due to inhibition of protein hydrolysis in the lysosomes and subsequent recycling of receptors [31, 35]. Not all proteins used in various experimental models inhibited bovine serum albumin endocytosis in mammalian PT cells [35–37]. When lysozyme was used instead of GFP in our frog experiments, it did not change the uptake of pre-injected YFP [19]. At the same time, 4-day lysozyme loading reduced YFP reabsorption and expression of endocytic receptors [19, 38]. Absorption capacity of PT cells was restored on the fifth day after cessation of loading and the number of YFP-associated profiles reached the control level [38]. Recovery of tubular YFP reabsorption occurred with a simultaneous increase in the number of internalized endocytic receptors and decrease in accumulation of lysozyme within PT cells (**Figure 6B**).

Thus, the results suggest the dependence of receptor-mediated endocytosis in the frog kidney on the molecular nature of absorbable ligands, conditions of their competitive absorption, and lysosomal accumulation in PT cells.

8. Conclusions

In general, morphophysiological study of the capacity for protein reabsorption in PT of the amphibian kidney was performed. Dose- and time-dependent tubular protein uptake and the existence of mechanisms limiting the protein absorption in epithelial PT cells were shown in frogs. Subcellular localization of endocytic receptors, megalin and cubilin, was revealed in amphibian PT cells after protein treatment. Intracellular trafficking of injected proteins was coincided with the distribution of megalin and cubilin. Specific marking of endocytic pathways revealed clathrin-dependent internalization of lysozymes and its subsequent transfer to endosomes. Thus, the protein uptake in the amphibian mesonephros is mediated by megalin and cubilin that confirms a critical role of endocytic receptors in the renal reabsorption of proteins in amphibians as in mammals. Based on our data, a frog model can be successfully used for investigating molecular mechanisms involved in the process of renal protein reabsorption and its comparative aspects.

Acknowledgements

This work was supported by the Russian Foundation for Basic Research (RFBR) (project no. 16-04-00465) for Sections 3 and 6–8 and state budget funding according to the assignment by the Russian Federal Agency for Scientific Organizations (FASO Russia) for Sections 1, 2, 4, and 5 (Physiological and biochemical mechanisms of homeostasis and their evolution).

Author details

Elena V. Seliverstova* and Natalya P. Prutskova

*Address all correspondence to: elena306@yandex.ru

Laboratory of Renal Physiology, Sechenov Institute of Evolutionary Physiology and Biochemistry of the Russian Academy of Sciences, St. Petersburg, Russia

References

[1] Christensen EI, Birn H, Storm T, Weyer TK, Nielsen R. Endocytic receptors in the renal proximal tubule. Physiology. 2012;**27**:223-236

[2] Christensen EI, Verroust PJ, Nielsen R. Receptor-mediated endocytosis in renal proximal tubule. Pflügers Archiv. 2009;**458**:1039-1048

[3] Gross ML, Hanke W, Koch A, Ziebart H, Amann K, Ritz E. Intraperitoneal protein injection in the axolotl: the amphibian kidney as a novel model to study tubulointerstitial activation. Kidney International. 2002;**62**:51-59

[4] Natochin YV. Evolutionary aspects of renal function. Kidney International. 1996; **49**:1539-1542

[5] Beyenbach KW. Comparative physiology of the renal proximal tubule. Renal Physiology. 1985;**8**:222-236

[6] Renkin EM, Gilmore JP. Glomerular filtration. In: Orloff J, Berliner RW, editors. Handbook of Physiology. Section 8: Renal Physiology. Baltimore: William and Wilkins Co; 1973. pp. 185-248

[7] Schaffner A, Rodewald R. Glomerular permeability in the bullfrog *Rana catesbeiana*. The Journal of Cell Biology. 1978;**79**:314-328

[8] Taugner R, Schiller A, Ntokalou-Knittel S. Cells and intercellular contacts in glomeruli and tubules of the frog kidney. A freeze-fracture and thin-section study. Cell and Tissue Research. 1982;**226**:589-608

[9] Fenoglio C, Vaccarone R, Chiari P, Gervaso MV. An ultrastructural and cytochemical study of the mesonephros of *Rana esculenta* during activity and hibernation. European Journal of Morphology. 1996;**34**:107-121

[10] Tanner GA, Rippe C, Shao Y, Evan AP, Williams JC Jr. Glomerular permeability to macromolecules in the Necturus kidney. American Journal of Physiology. Renal Physiology. 2009;**296**:F1269-F1278

[11] Richter S, Splechtna H. The structure of anuran podocyte—Determined by ecology? Acta Zoologica (Stockholm). 1996;**77**:335-348

[12] Meseguer J, García-Ayala A, López-Ruiz A, Esteban MA. Structure of the amphibian mesonephric tubule during ontogenesis in *Rana ridibunda* L. tadpoles: Early ontogenetic stages, renal corpuscle formation, neck segment and peritoneal funnels. Anat. Embryol. (Berl). 1996;**193**:397-406

[13] Leheste JR, Rolinski B, Vorum H, Hilpert J, Nykjaer A, Jacobsen C, Aucouturier P, Moskaug JO, Otto A, Christensen EI, Willnow TE. Megalin knockout mice as an animal model of low molecular weight proteinuria. The American Journal of Pathology. 1999; **155**:361-1370

[14] Zhou X, Vize PD. Proximo-distal specialization of epithelial transport processes within the Xenopus pronephric kidney tubules. Developmental Biology. 2004;**271**:322-338

[15] Prutskova NP, Seliverstova EV. Tubular GFP uptake pattern in the rat and frog kidneys. Comparative Biochemistry and Physiology. Part A, Molecular & Integrative Physiology. 2011;**160**:175-183

[16] Seliverstova EV, Burmakin MV, Natochin Yu V. Renal clearance of absorbed intact GFP in the frog and rat intestine. Comparative Biochemistry and Physiology. Part A. 2007; **147**:1067-1073

[17] Seliverstova EV, Prutskova NP. Receptor-mediated endocytosis of lysozyme in renal proximal tubules of the frog *Rana temporaria*. European Journal of Histochemistry. 2015;**59**:79-86

[18] Prutskova NP, Seliverstova EV. Absorption capacity of renal proximal tubular cells studied by combined injections of YFP and GFP in *Rana temporaria* L. Comparative Biochemistry and Physiology, Part A. 2013;**166**:138-146

[19] Prutskova NP, Seliverstova EV. Effect of different proteins on reabsorption of yellow fluorescent protein in the kidney of *Rana temporaria*. Journal of Evolutionary Biochemistry and Physiology. 2016;**52**:221-225

[20] Prutskova NP, Kutina AV. Comparative analysis of the fluorescent protein reabsorption in frog and rat kidneys. Journal of Evolutionary Biochemistry and Physiology. 2015;**51**: 221-224

[21] Seliverstova EV, Prutskova NP. Reabsorption of yellow fluorescent protein in the *Rana temporaria* kidney by receptor-mediated endocytosis. Journal of Evolutionary Biochemistry and Physiology. 2014;**50**:522-530

[22] Boutilier RG, Stiffler DF, Toews DP. Exchange of respiratory gases, ions, and water in amphibious and aquatic amphibians. In: Feder ME, Burggren WW, editors. Environmental Physiology of the Amphibians. The University of Chicago: The University of Chicago Press; 1992. pp. 81-116

[23] Ginetsinskii AG. Physiological Mechanisms of Fluid Electrolyte Homeostasis (in Russian). Leningrad: Nauka; 1964. 428 p

[24] Braun ED, Dantzler WH. Vertebrate renal system. In: Dantzler WH, editor. Handbook of Physiology, Section 13, Comparative Physiology. Vol. I. New York: Oxford University Press; 1997. pp. 481-576

[25] SD MC, Bradshaw D. Hormonal control of salt and water balance in vertebrates. General and Comparative Endocrinology. 2006;**147**:3-8

[26] Pang PK, Galli-Gallardo SM, Collie N, Sawyer WH. Renal and peripheral vascular responsiveness to arginine vasotocin in bullfrog, *Rana catesbeiana*. The American Journal of Physiology. 1980;**239**:R156-R160

[27] Boyd SK. Amphibian neurohypophysial peptides. In: Kastin AJ, editor. The Handbook of Biologically Active Peptides, Chapter 49. 1st ed. San Diego: Academic Press; 2006. pp. 327-332

[28] Uchiyama M, Konno N. Hormonal regulation of ion and water transport in anuran amphibians. General and Comparative Endocrinology. 2006;**147**:54-61

[29] Ammar A, Rajerison RM, Roseau S, Bloch-Faure M, Butlen D. Frog glomerular vasotocin receptors resemble mammalian V1b receptors. The American Journal of Physiology. 1994;**267**:R1198-R1208

[30] Prutskova NP. Renal filtration and reabsorption of GFP in *Rana temporaria*: effect of arginine-vasotocin. Journal of Evolutionary Biochemistry and Physiology. 2011;**47**:54-61

[31] De S, Kuwahara S, Saito A. The endocytic receptor megalin and its associated proteins in proximal tubule epithelial cells. Membranes (Basel). 2014;**3**:333-355

[32] Christensen E, Raciti D, Reggiani L, Verroust PJ, Brändli AW. Gene expression analysis defines the proximal tubule as the compartment for endocytic receptor-mediated uptake in the Xenopus pronephric kidney. Pflügers Archiv. 2008;**456**:1163-1176

[33] Anzenberger U, Bit-Avragim N, Rohr S, Rudolph F, Dehmel B, Willnow TE, Abdelilah-Seyfried S. Elucidation of megalin/LRP2-dependent endocytic transport processes in the larval zebrafish pronephros. Journal of Cell Science. 2006;**119**:2127-2137

[34] Brodsky FM, Chen CY, Knuehl C, Towler MC, Wakeham DE. Biological basket weaving: formation and function of clathrin-coated vesicles. Annual Review of Cell and Developmental Biology. 2001;**7**:517-568

[35] Lee D, Gleich K, Fraser SA, Katerelos M, Mount PF, Power DA. Limited capacity of proximal tubular proteolysis in mice with proteinuria. American Journal of Physiology. Renal Physiology. 2013;**304**:F1009-F1019

[36] Gekle M, Mildenberger S, Freudinger R, Silbernagl S. Long-term protein exposure reduces albumin binding and uptake in proximal tubule-derived opossum kidney cells. Journal of the American Society of Nephrology. 1998;**9**:960-968

[37] Zhai XY, Nielsen R, Birn H, Drumm K, Mildenberger S, Freudinger R, Moestrup SK, Verroust PJ, Christensen EI, Gekle M. Cubilin and megalin-mediated uptake of albumin in cultures proximal tubule cells of opossum kidney. Kidney International. 2000; **58**:1523-1533

[38] Seliverstova EV, Prutskova NP. Tubular protein uptake pattern in the frog model (*Rana temporaria*) the effect of previous protein loading. Journal of Evolutionary Biochemistry and Physiology. 2017;**53**:215-224

Chapter 12

The Contribution of Changes in Adenylyl Cyclase Signaling System of the Brain and Myocardium to Etiology and Pathogenesis of Diabetes Mellitus

Alexander O. Shpakov

Additional information is available at the end of the chapter

http://dx.doi.org/10.5772/intechopen.73661

Abstract

The functional changes in hormone-sensitive adenylyl cyclase (AC) signaling system of the central nervous system (CNS) and periphery play a crucial role in etiology and pathogenesis of diabetes mellitus (DM). The identification of these changes in AC signaling system and the abnormalities in AC signaling network are necessary for creation of the new strategies to treat and prevent diabetic pathology. In this chapter, our data and the results of other authors on the changes in hormone-sensitive adenylyl cyclase signaling system (ACSS) in the diabetic brain and heart and on their contribution to etiology and pathogenesis of DM and its complications, diabetic cardiomyopathy in particular, are presented and analyzed, and the promising approaches to treat DM and its complications, which are based on the restoration of AC signaling cascades and their functional interaction, are discussed.

Keywords: diabetes mellitus, adenylyl cyclase system, brain, myocardium, bromocriptine

1. Introduction

Diabetes mellitus (DM) is a major global health problem affecting more than 350 million people worldwide. It is one of the most severe metabolic disorders in humans characterized by hyperglycemia due to insulin deficiency or insulin resistance of target tissues. Insulin-dependent, type 1, and non-insulin-dependent, type 2, DM (DM1 and DM2) induce a large number of diseases in the nervous, cardiovascular, endocrine, and other systems, and these complications of DM are found in more than one-quarter of diabetic patients [1–4]. It is generally accepted that the changes in hormonal signaling systems in the CNS and periphery play a crucial role in etiology and pathogenesis of DM and its complications. For a long time, the main attention was focused on the signaling systems regulated by insulin, insulin-like growth factor-1, and leptin, whose functional activity is largely impaired in DM1 and DM2 [5–8]. Meanwhile, in recent years, numerous data in favor of the close relationship between the changes in G protein-coupled signaling systems and the pathogenesis of DM were obtained. These systems are regulated by a broad spectrum of hormonal agents, such as amino acids and their derivatives, peptide and glycoprotein hormones, and nucleotides, which bind specifically to G protein-coupled receptors (GPCRs) seven times penetrating the plasma membrane.

The central role among these systems belongs to adenylyl cyclase signaling system (ACSS), which is represented in all types of cells and tissues and is responsible for hormonal regulation of fundamental cellular processes. The ACSS has the following main components: (1) G protein-coupled receptor (GPCR) recognizing and specifically interacting with hormonal stimuli, (2) $\alpha\beta\gamma$-heterotrimeric G protein of the stimulatory (G_s) and inhibitory (G_i) types, (3) the enzyme adenylyl cyclase (AC) catalyzing the formation of cyclic AMP (cAMP), and (4) cAMP-activated protein kinase (PKA) and cAMP-activated guanine nucleotide exchange factors (Epac1 and Epac2) that control the cAMP-dependent intracellular cascades and transcription factors. As the pathological changes in ACSS lead to dysfunctions in most organs and tissues, they are one of the causes of severe complications of DM such as diabetic cardiomyopathy, nephropathy, encephalopathy, and metabolic and endocrine disorders.

This chapter describes our data and the results of other authors on the changes and abnormalities in hormone-sensitive ACSS in the diabetic brain and heart and their contribution into etiology and pathogenesis of DM and its complications, diabetic cardiomyopathy in particular, and on the approaches to treat DM, which are based on the restoration of AC-signaling cascades and their functional interaction.

2. The ACSS in the diabetic brain

It is shown that in DM, the functional activity of cAMP-dependent signaling pathways regulated by dopamine (DA), serotonin, and melanocortin peptides in the brain and especially in its hypothalamic area is changed significantly. This triggers neurodegenerative processes in the CNS and affects the central regulation of energy homeostasis, inducing peripheral insulin resistance and abnormalities in the lipid and carbohydrate metabolism (**Figure 1**).

The brain DA controls locomotor activity, cognition, feeding behavior, and via central mechanisms regulates functions of the endocrine and cardiovascular systems. The DA stimulates AC activity via G_s protein-coupled dopamine receptor of the type 1 (DA_1R) and inhibits hormone-stimulated AC activity via G_i protein-coupled DA_2R. In the hypothalamus and brainstem of rats with the streptozotocin (STZ) model of DM1, the concentration of DA and the number of DA_2R decreased significantly [9]. The hypothalamus and brainstem are involved in the control of glucose homeostasis and feeding behavior. In patients and experimental animals with DM2 and metabolic syndrome, the activity of brain D_2-dopaminergic system also reduced, as illustrated by a decrease of dopamine level and DA_2R expression [9, 10].

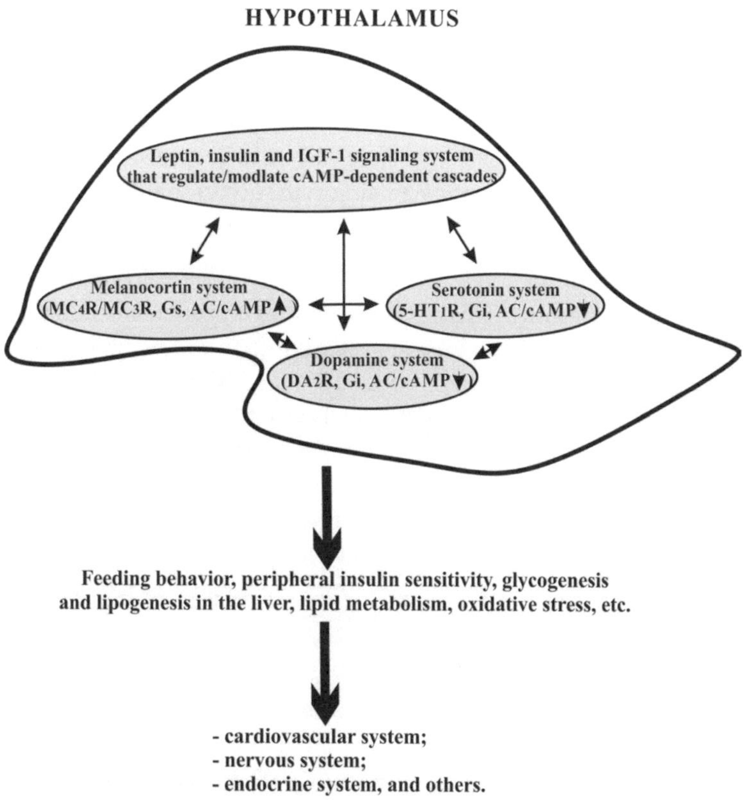

Figure 1. The interaction between hypothalamic signaling systems and their influence on the peripheral energy homeostasis and the functioning of organs and tissues.

We showed that in the brain and in the hypothalamus of rats with acute and moderate STZ-induced DM1 and with neonatal and high fat diet (HFD)/STZ models of DM2, the inhibitory effect of DA_2R-agonists on forskolin-stimulated AC activity and on the expression of the *Drd2* gene encoding DA_2R was reduced significantly, especially in DM1 [11–14]. Meanwhile, the functional activity of DA_1R pathway in the CNS of diabetic rats was changed to a small extent.

The restoration of brain D_2-dopaminergic system in experimental and human DM2 can be achieved using the alkaloid bromocriptine (BC), a selective DA_2R-agonist that activates DA_2R and decreases intracellular cAMP level in neurons. In DM2, the BC inhibits the activity of hypothalamic neurons controlling glucose production and lipid synthesis in the liver, activates dopaminergic neurons regulating insulin sensitivity, and via central mechanisms improves functions of the cardiovascular system, preventing the development of severe forms of diabetic cardiomyopathy [10, 15, 16]. The effect of BC therapy on glucose homeostasis in DM2 is comparable to that of metformin, widely used antidiabetic drug, and, as demonstrated in clinical trials and in animal models, the co-administration of BC with metformin, glipizide, and pioglitazone enhances their glucose-lowering effect and reduces effective doses of these drugs, thereby preventing their adverse effects [17, 18]. The glucose-lowering effect of glipizide when co-administered with BC is also increased in rats with the alloxan model of DM1 [19].

We demonstrated that 2-month BC treatment of rats with HFD-induced DM2 resulted in the improved glucose homeostasis and insulin sensitivity [20, 21]. The BC partially restored sensitivity of brain ACSS to agonists of 5-hydroxytryptamine receptors of the subtype 1B (5-$HT_{1B}R$) and somatostatin receptors, which indicates a functional relationship between DA_2R signaling and the somatostatin and serotonin systems in the CNS [21]. The treatment of diabetic rats with BC also led to normalization of adrenergic signaling in the myocardium and to restoration of AC sensitivity to gonadotropin in testes, indicating a broad therapeutic potential of BC in DM2 [20].

Brain serotonin, acting on different types of 5-HTRs, regulates feeding behavior, motor activity, pain, depression, and learning. This neurotransmitter is also involved in the control of the cardiovascular, endocrine, and reproductive systems and in the regulation of production of insulin and other hormones by pancreatic islets [22]. It was shown that in patients with DM1 and DM2 and in animals with experimental models of DM1, the brain level of serotonin and its precursor tryptophan and the ratio of free-total tryptophan were significantly decreased. The decreased serotonin level and the changes in serotonin metabolism due to decrease in activity of tryptophan-5-hydroxylase-2, the rate-limiting enzyme in serotonin biosynthesis, led to impairment of serotonin signaling pathways in the brain and to alteration of the number and affinity of 5-HTRs, which weakens serotonin-mediated regulation of lipid and carbohydrate metabolism and insulin sensitivity [23, 24].

Based on serotonin deficiency in the diabetic brain, it can be assumed that increasing the serotonin level in CNS is an appropriate approach to normalize feeding behavior and improve glucose homeostasis and insulin sensitivity impaired in diabetic pathology. This suggestion is confirmed by the results obtained in treating diabetic patients with fluoxetine and other selective serotonin reuptake inhibitors. These inhibitors induced weight loss, reduced the plasma levels of glucose and glycated hemoglobin, and improved insulin sensitivity [25, 26].

We showed that long-term treatment of rats with neonatal and HFD/STZ DM2 using intranasally administered serotonin (IS) restored hormonal sensitivity of ACSS in the brain and in the periphery and improved metabolic parameters and cognitive functions [12, 13, 27]. The 8-week treatment of female rats with neonatal DM2 with IS (20 μg/rat daily) restored AC-mediated regulatory effects of monoamines and relaxin in the brain, β-adrenergic agonists in the myocardium, and gonadotropins in ovaries [12, 28]. Along with it, using the Morris water test, we found that IS treatment improved DM-induced impairment of learning and spatial memory [12]. The 2-month IS treatment of male rats with HFD/low-dose STZ model of DM2 decreased the body weight, improved the glucose tolerance and insulin-induced glucose utilization, and also reduced the level of triglycerides and LDL-cholesterol, and the LDL/HDL-cholesterol ratio, which indicates the normalization of lipid metabolism. Besides, IS treatment restored hormonal sensitivity of ACSS in the hypothalamus and normalized the ratio of β_1-, β_2-, and β_3-adrenergic receptors (β-ARs) in the myocardium of diabetic rats. Based on these findings, we can conclude that increasing the brain serotonin level may be an effective way to treat DM2 and its complications that are induced by abnormalities in the brain and peripheral AC signaling [13].

The hypothalamic melanocortin system plays a very important role in regulation of feeding behavior, insulin sensitivity, and lipid metabolism [29]. The sensor components of this system are G_s protein-coupled melanocortin receptors of the types 3 and 4 (MC_3R and MC_4R). They are activated by α-melanocyte-stimulating hormone (α-MSH) and other peptides of the melanocortin family generated from pro-opiomelanocortin (POMC) that is produced by POMC-expressing neurons of the arcuate nucleus of hypothalamus. The binding of MC_3R and MC_4R with agonists leads to activation of AC and cAMP-dependent signaling cascades. Along with melanocortin peptides, the agouti-related peptide (AgRP) with MC_4R antagonistic activity is produced in the arcuate nucleus, and it inhibits regulatory effects of α-MSH and triggers G protein-independent arrestin signaling [30].

The inhibition of MC_4R-signaling cascades led to hyperphagia, metabolic disorders, insulin resistance, and eventually to DM2 [31]. Mice lacking MC_4R and agouti mice with increased AgRP expression had the reduced insulin sensitivity, and treatment of healthy mice by MC_4R antagonists and high-dose AgRP enhanced appetite and induced insulin resistance [32, 33]. Some patients with DM2 and metabolic syndrome were characterized by mutations in the *Mc4r* gene and by impaired MC_4R signaling [34]. The α-MSH and other MC_4R agonists, on the contrary, had an antidiabetic effect when administered to rodents with obesity and insulin resistance. They reduced food intake and normalized the glucose and insulin level and energy metabolism [31, 32].

We showed that in the hypothalamus of rats with neonatal and HFD/low-dose STZ DM2, the activity of MC_4R-signaling pathway decreased significantly. This was illustrated by decrease of the *Mc4r* gene expression and the stimulating effects of α-MSH and selective MC_4R-agonist THIQ on AC activity and GTP-binding capacity of G_s proteins. The long-term treatment of diabetic animals with BC and intranasally administered insulin and serotonin significantly restored MC_4R signaling, and this coincided with the improvement of insulin sensitivity and the carbohydrate and lipid metabolism [13].

One of the approaches to restore hypothalamic melanocortin system in DM2 is the use of selective MC_4R-agonists, as demonstrated in the experiments with obese and diabetic animals. But currently, there are no available highly selective MC_4R-agonists, while melanotan-II, the most widely used non-selective MCR agonist, leads to a large number of adverse effects [35]. Currently, new agonists of MC_4R are being developed intensively, but they have not been used in clinic yet. The most effective among them are α-MSH analogs modified by fatty acid radicals at the N-terminus, highly selective MC_4R agonist BIM-22493 [36, 37]. The BIM-22493 easily penetrates across blood-brain barrier, activates hypothalamic MC_4R-signaling pathways, and, as a result, decreases food intake, body weight and fat mass, and improves glucose tolerance. It should be noted that even a long-term treatment of experimental animals with BIM-22493 had no adverse effects on the cardiovascular system and blood pressure [37]. The effectiveness of MC_4R agonists can be significantly enhanced when they are combined with agonist of glucagon-like peptide-1 receptor, which is widely used to treat DM2. Co-administration of BIM-22493 and liraglutide, a stable agonist of glucagon-like peptide-1 receptor, into diabetic mice prevented insulin resistance and improved energy expenditure much more effectively as compared to monotherapy [38].

3. The ACSS in the diabetic heart

The DM1 and DM2 are closely associated with severe cardiovascular diseases, such as acute myocardial infarction, congestive heart failure, and atherosclerosis [39, 40]. The pathological changes in contractile function of the heart in DM are largely due to impairment of the adrenergic, cholinergic, and purinergic pathways of AC regulation in cardiomyocytes [28, 39, 41–44]. The adrenergic signaling has a very important role in the functioning of the cardiovascular system, and it changes to the greatest extent in DM. In rats with the STZ model of DM1, the expression of genes encoding β-ARs and the activity of the receptors are altered and the pathological changes are enhanced with increasing duration and severity of DM [41]. In the cardiac muscle, there are three pharmacologically distinct subtypes of β-ARs. The G_s protein-coupled β_1-AR stimulates AC activity, β_2-AR interacts with the G_s and G_i proteins, and is able to both stimulate and inhibit AC activity, while β_3-AR interacts preferably with G_i protein, inhibiting AC.

In diabetic rats with 6–14 week DM1, the expression of gene encoding β_1-AR was significantly reduced, while the expression of β_2-AR gene, on the contrary, was increased. The number of functionally active β-ARs on the surface of cardiomyocytes was reduced for both β_1- and β_2-ARs, which is caused by increasing the rate of β_2-AR degradation and the deterioration of post-translational processing of receptor [41, 45]. Meanwhile, the mRNA level for β_3-AR and the number of these receptors on the surface of cardiomyocytes in rats with 14-weeks DM1 increased 2 or more times in comparison with control animals. The specificity of changes for β-AR subtypes in diabetic heart resulted in alteration of the $\beta_1/\beta_2/\beta_3$ ratio. In the myocardium of diabetic rats, the ratio was 40:36:23, while in the myocardium of healthy rats, the ratio was 62:30:8 [45]. The treatment of diabetic rats with insulin led to normalization of the $\beta_1/\beta_2/\beta_3$ ratio (57:33:10). The specific changes in β-AR activity including two or threefold increase in the number of β_3-AR were identified in the cardiac muscle of patients with DM2 and metabolic syndrome, as well as in patients with acute heart failure [46]. The study of genotype of patients with DM2 and metabolic syndrome allowed detecting the mutation in a codon 64 of β_3-AR gene, which led to a significant increase of activity of mutant receptor [47]. We also showed significant changes in the $\beta_1/\beta_2/\beta_3$ ratio in the myocardium of rats with different models of diabetic pathology, and the ratio was restored when the animals were treated with intranasal insulin and, in the case of DM2, with D_2-agonist BC and metformin [13, 20, 28, 44]. The increase of β_3-AR activity prevents AC hyperactivation caused by the increased catecholamine levels characteristic for diabetic cardiomyopathy. The increase of β_3-AR activity can also be a compensatory mechanism contributing to the preservation of functional activity of endothelial NO-synthase and soluble guanylyl cyclase that regulate vascular contractility [46]. However, with prolonged duration of DM, the increase of β_3-AR signaling in the myocardium leads to imbalance of adrenergic regulation and induces the negative inotropic effect of β-AR agonists and bradycardia [41].

The apoptotic processes in the cardiac muscle contribute significantly to etiology and pathogenesis of diabetic cardiomyopathy, and they largely depend on the β_1-AR signaling. A decrease in β_1-AR activity in the cardiac muscle in DM1 leads to inhibition of apoptotic processes in cardiomyocytes and prevents myocardial dysfunction and acute heart failure. It should be noted that in healthy animals, β_1-AR agonists induce apoptosis in rat cardiomyocytes, while β_1-AR antagonists suppress it [48].

We studied ACSS activity in the myocardium of rats with acute DM1 induced by high-dose STZ and found the decrease of the basal level of GTP-binding and the AC stimulating effect of guanine nucleotides, which indicates a weakening of G_s protein function in cardiomyocytes of diabetic animals [49]. Meanwhile, the stimulation of AC by forskolin that directly interacts with catalytic site of the enzyme did not change, indicating the preservation of AC catalytic activity. The AC stimulating effect of β-agonists was decreased, but to a small extent, while the corresponding effect of relaxin, a peptide hormone that plays an important role in regulation of the cardiovascular system, was decreased by 48%. The study of ACSS in the heart of rats with 7-month DM1 induced by multiple injections of low-dose STZ shows the decrease of both basal and forskolin/guanine nucleotides-stimulated AC activity, demonstrating the reduced activity of both AC and G_s protein [50]. A significant decrease in the norepinephrine and isoproterenol effects on AC activity, more pronounced than in acute DM1, was also observed. The changes of ACSS activity significantly depended on the age of rats when DM1 was initiated [44, 50]. Our results indicate that changes in adrenergic signaling cascades in the heart are highly dependent on the experimental model of DM1.

Unlike DM1, in DM2, the number of β-ARs in the myocardium is not substantially different from control, but the sensitivity of β-ARs to agonists and their effects on AC are decreased significantly [11, 13, 51–53]. The changes in β-AR signaling strongly vary in rats with different models of DM2 depending on duration and severity of the disease [11, 44, 49, 53, 54]. In the myocardium of rats with 8-month neonatal model of DM2, the effect of isoproterenol on AC activity was increased, although to a small extent. When the duration of DM2 was 18 months, this effect was reduced as compared with the control group. The stimulating effect of relaxin on AC was reduced in DM2 with different durations, and in 18-months DM2, it did not exceed 46% of that in control [42]. The decrease of effect of guanine nucleotides on AC was shown, indicating a weakening of G_s protein function, and one of the causes for this is hyperhomocysteinemia, typical for severe DM2 [54].

It was shown that the treatment of diabetic rats with thyroid hormone levothyroxine was effective for restoration of the number and functional activity of β-ARs [55, 56]. This indicates a close relationship between hypothyroid state, typical for human DM1 and DM2, and impaired myocardial function in DM. In this regard, there are serious grounds to believe that one of the approaches to prevent diabetic cardiomyopathy is restoration of hypothalamic–pituitary-thyroid axis and compensation of thyroid hormones deficiency. The treatment of DM2 rats with D_2-agonist BC and intranasally administered insulin and serotonin, restoring hypothalamic ACSS, also improves the function of the cardiovascular system and sensitivity of myocardial AC to hormonal regulators [20, 28].

4. Concluding remarks

Summing up, the changes in hormone-regulated ACSS in the brain and heart and abnormalities of interaction between them are the most important factors leading to the development of DM and its complications. Consequently, the identification of disturbances in these cascades and the development of approaches to their correction should be regarded as the most promising strategy to treat and prevent diabetic pathology. The causal link between DM and the pathological changes in AC signaling is not a one-way avenue, from DM to these changes in the organs and tissues and, further, to diabetic encephalopathy, cardiomyopathy, and other complications of DM. The opposite situation can also be realized when impaired AC signaling triggers the processes leading to DM. The dysfunctions in the brain ACSS sensitive to melanocortin peptides and monoamines can induce DM2 and metabolic syndrome, while dysregulation of cAMP signaling in the pancreatic islets weakens insulin-producing function of β-cells and provokes the development of DM1. This speaks in favor of the use of a wide scale of hormonal and non-hormonal agents that control AC activity and influence the availability, transport, and secretion of hormonal molecules in the treatment and prevention of DM. The development of new approaches for the treatment of DM, which are based on the monitoring and correction of the ACSS activity in the brain, myocardium, and the other organs and tissues, requires a detailed study of the changes in the ACSS in different forms of experimental and human DM, as well as the effects on the ACSS of a number of the factors, such as the duration and severity of DM, the DM treatment with insulin, metformin, and other drugs, the frequency of hypoglycemic episodes, and the DM-induced complications. Nowadays, in our Laboratory of Molecular Endocrinology and Neurochemistry, Sechenov Institute of Evolutionary Physiology and Biochemistry, we use a lot of models of DM and various approaches of molecular endocrinology, pharmacology, and experimental medicine in order to understand etiology and pathogenesis of DM and its complications and to propose the new strategies to treat them.

Acknowledgements

This work was supported by the Russian Science Foundation (No 14–15-00413) (Section 2) and by the state assignment of FASO of Russia ("The mechanisms of development of neuropsychic, metabolic and hormonal dysfunctions in the nervous and endocrine diseases and the approaches for their correction") (Section 3).

Author details

Alexander Olegovich Shpakov

Address all correspondence to: alex_shpakov@list.ru

I.M. Sechenov Institute of Evolutionary Physiology and Biochemistry RAS, St. Petersburg, Russia

References

[1] Tabit CE, Chung WB, Hamburg NM, Vita JA. Endothelial dysfunction in diabetes mellitus: Molecular mechanisms and clinical implications. Reviews in Endocrine & Metabolic Disorders. 2010;**11**:61-74. DOI: 10.1007/s11154-010-9134-4

[2] Magge SN. Cardiovascular risk in children and adolescents with type 1 and type 2 diabetes mellitus. Current Cardiovascular Risk Reports. 2012;**6**:591-600. DOI: 10.1007/s12170-012-0274-9

[3] Umegaki H, Hayashi T, Nomura H, Yanagawa M, Nonogaki Z, Nakshima H, et al. Cognitive dysfunction: An emerging concept of a new diabetic complication in the elderly. Geriatrics & Gerontology International. 2013;**13**:28-34. DOI: 10.1111/j.1447-0594.2012.00922.x

[4] Kahaly GJ, Hansen MP. Type 1 diabetes associated autoimmunity. Autoimmunity Reviews. 2016;**15**:644-648. DOI: 10.1016/j.autrev.2016.02.017

[5] Clemmons DR. Metabolic actions of insulin-like growth factor-I in normal physiology and diabetes. Endocrinology and Metabolism Clinics of North America. 2012;**41**:425-443. DOI: 10.1016/j.ecl.2012.04.017

[6] Meek TH, Morton GJ. Leptin, diabetes, and the brain. Indian Journal of Endocrinology and Metabolism. 2012;**16**:534-542. DOI: 10.4103/2230-8210.105568

[7] Shpakov AO, Derkach KV, Berstein LM. Brain signaling systems in the type 2 diabetes and metabolic syndrome: Promising target to treat and prevent these diseases. Future Science OA (FSO). 2015;**1**:FSO25. DOI: 10.4155/fso.15.23

[8] Shpakov AO. The brain leptin signaling system and its functional state in metabolic syndrome and type 2 diabetes mellitus. Journal of Evolutionary Biochemistry and Physiology. 2016;**52**:177-195. DOI: 10.1134/S0022093016030017

[9] Shankar PN, Joseph A, Paulose CS. Decreased [^{3}H] YM-09151-2 binding to dopamine D2 receptors in the hypothalamus, brainstem and pancreatic islets of streptozotocin-induced diabetic rats. European Journal of Pharmacology. 2007;**557**:99-105. DOI: 10.1016/j.ejphar.2006.11.018

[10] Grunberger G. Novel therapies for the management of type 2 diabetes mellitus: Part 1. Pramlintide and bromocriptine-QR. Journal of Diabetes. 2013;**5**:110-117. DOI: 10.1111/1753-0407.12034

[11] Shpakov AO, Chistyakova OV, Derkach KV, Moiseyuk IV, Bondareva VM. Intranasal insulin affects adenylyl cyclase system in rat tissues in neonatal diabetes. Central European Journal of Biology. 2012;**7**:33-47. DOI: 10.2478/s11535-011-0089-6

[12] Shpakov AO, Derkach KV, Chistyakova OV, Sukhov IB, Shipilov VN, Bondareva VM. The brain adenylyl cyclase signaling system and cognitive functions in rat with neonatal diabetes under the influence of intranasal serotonin. Journal of Metabolic Syndrome. 2012;**1**:1000104. DOI: 10.4172/jms.1000104

[13] Derkach KV, Bondareva VM, Chistyakova OV, Berstein LM, Shpakov AO, The effect of long-term intranasal serotonin treatment on metabolic parameters and hormonal signaling in rats with high-fat diet/low-dose streptozotocin-induced type 2 diabetes. International Journal of Endocrinology. 2015;**2015**:245459. DOI: 10.1155/2015/245459

[14] Sukhov IB, Derkach KV, Shpakov AO. Comparative study of functional activity of the D2-dopaminergic system in the hypothalamus of rats with different models of diabetes mellitus. Journal of Evolutionary Biochemistry and Physiology. 2016;**52**:267-269. DOI: 10.1134/S0022093016030108

[15] Scranton R, Cincotta A. Bromocriptine—Unique formulation of a dopamine agonist for the treatment of type 2 diabetes. Expert Opinion on Pharmacotherapy. 2010;**11**:269-279. DOI: 10.1517/14656560903501544

[16] Weiland CM, Hilaire ML. Bromocriptine mesylate (Cycloset) for type 2 diabetes mellitus. American Family Physician. 2013;**87**:718-720 http://www.aafp.org/afp/steps

[17] Vinik AI, Cincotta AH, Scranton RE, Bohannon N, Ezrokhi M, Gaziano JM. Effect of bromocriptine-QR on glycemic control in subjects with uncontrolled hyperglycemia on one or two oral anti-diabetes agents. Endocrine Practice. 2012;**18**:931-943. DOI: 10.4158/EP12187.OR

[18] Ghosh A, Sengupta N, Sahana P, Giri D, Sengupta P, Das N. Efficacy and safety of add on therapy of bromocriptine with metformin in Indian patients with type 2 diabetes mellitus: A randomized open labeled phase IV clinical trial. Indian Journal of Pharmacology. 2014;**46**:24-28. DOI: 10.4103/0253-7613.125160

[19] Kumar VSH, Vinutha MB, Aithal S, Baleed SR, Patil UN. Bromocriptine, a dopamine (d2) receptor agonist, used alone and in combination with glipizide in sub-therapeutic doses to ameliorate hyperglycaemia. Journal of Clinical and Diagnostic Research. 2013;**7**:1904-1907. DOI: 10.7860/JCDR/2013/5803.3347

[20] Derkach KV, Bondareva VM, Moiseyuk IV, Shpakov AO. The effect of 2-month bromocriptine treatment on the activity of the adenylate cyclase signaling system in the myocardium and testes of rats with type 2 diabetes. Cell and Tissue Biology. 2015;**9**:395-405. DOI: 10.1134/S1990519X15050041

[21] Shpakov AO, Derkach KV, Chistyakova OV, Bondareva VM. The influence of bromocryptine treatment on activity of the adenylyl cyclase system in the brain of rats with type 2 diabetes mellitus induced by high-fat diet. Doklady Biochemistry and Biophysics. 2014;**459**:186-189. DOI: 10.1134/S1607672914060027

[22] Barnes NM, Sharp T. A review of central 5-HT receptors and their function. Neuropharmacology. 1999;**38**:1083-1152. DOI: 10.1016/S0028-3908(99)00010-6

[23] Kloiber S, Kohli MA, Brueckl T, Ripke S, Ising M, Uhr M, et al. Variations in tryptophan hydroxylase 2 linked to decreased serotonergic activity are associated with elevated risk for metabolic syndrome in depression. Molecular Psychiatry. 2010;**15**:736-747. DOI: 10.1038/mp.2008.142

[24] Herrera-Marquez R, Hernandez-Rodriguez J, Medina-Serrano J, Boyzo-Montes de Oca A, Manjarrez-Gutierrez G. Association of metabolic syndrome with reduced central serotonergic activity. Metabolic Brain Disease. 2011;**26**:29-35. DOI: 10.1007/s11011-010-9229-3

[25] Goodnick PJ. Use of antidepressants in treatment of comorbid diabetes mellitus and depression as well as in diabetic neuropathy. Annals of Clinical Psychiatry. 2001;**13**:31-41. PMID: 11465683

[26] Deuschle M. Effects of antidepressants on glucose metabolism and diabetes mellitus type 2 in adults. Current Opinion in Psychiatry. 2013;**26**:60-65. DOI: 10.1097/YCO.0b013e32835a4206

[27] Sukhov IB, Derkach KV, Chistyakova OV, Bondareva VM, Shpakov AO. The effect of prolonged intranasal administration of serotonin on the activity of hypothalamic signaling systems in male rats with neonatal diabetes. Cell and Tissue Biology. 2016;**10**:314-323. DOI: 10.1134/S1990519X1604012X

[28] Shpakov AO, Derkach KV, Chistyakova OV, Moiseyuk IV, Sukhov IB, Bondareva VM. Effect of intranasal insulin and serotonin on functional activity of the adenylyl cyclase system in myocardium, ovary, and uterus of rats with prolonged neonatal model of diabetes mellitus. Journal of Evolutionary Biochemistry and Physiology. 2013;**49**:153-164. DOI: 10.1134/S0022093013020047

[29] Jeong JK, Kim JG, Lee BJ. Participation of the central melanocortin system in metabolic regulation and energy homeostasis. Cellular and Molecular Life Sciences. 2014;**71**:3799-3809. DOI: 10.1007/s00018-014-1650-z

[30] Bäckberg M, Madjid N, Ogren SO, Meister B. Down-regulated expression of agouti-related protein (AGRP) mRNA in the hypothalamic arcuate nucleus of hyperphagic and obese tub/tub mice. Brain Research. Molecular Brain Research. 2004;**125**:129-139. DOI: 10.1016/j.molbrainres.2004.03.012

[31] Nogueiras R, Wiedmer P, Perez-Tilve D, Veyrat-Durebex C, Keogh JM, Sutton GM, et al. The central melanocortin system directly controls peripheral lipid metabolism. The Journal of Clinical Investigation. 2007;**117**:3475-3488. DOI: 10.1172/JCI31743

[32] Balthasar N, Dalgaard LT, Lee CE, Yu J, Funahashi H, Williams T, et al. Divergence of melanocortin pathways in the control of food intake and energy expenditure. Cell. 2005;**123**:493-505. DOI: 10.1016/j.cell.2005.08.035

[33] Haskell-Luevano C, Schaub JW, Andreasen A, Haskell KR, Moore MC, Koerper LM, et al. Voluntary exercise prevents the obese and diabetic metabolic syndrome of the melanocortin-4 receptor knockout mouse. The FASEB Journal. 2009;**23**:642-655. DOI: 10.1096/fj.08-109686

[34] Pan Q, Delahanty LM, Jablonski KA, Knowler WC, Kahn SE, Florez JC, et al. Variation at the melanocortin 4 receptor gene and response to weight-loss interventions in the diabetes prevention program. Obesity (Silver Spring). 2013;**21**:520-526. DOI: 10.1002/oby.20459

[35] Wikberg JE, Mutulis F. Targeting melanocortin receptors: An approach to treat weight disorders and sexual dysfunction. Nature Reviews. Drug Discovery. 2008;**7**:307-323. DOI: 10.1038/nrd2331

[36] Conde-Frieboes K, Thøgersen H, Lau JF, Sensfuss U, Hansen TK, Christensen L, et al. Identification and in vivo and in vitro characterization of long acting and melanocortin 4 receptor (MC4-R) selective α-melanocyte-stimulating hormone (α-MSH) analogues. Journal of Medicinal Chemistry. 2012;**55**:1969-1977. DOI: 10.1021/jm201489a

[37] Kievit P, Halem H, Marks DL, Dong JZ, Glavas MM, Sinnayah P, et al. Chronic treatment with a melanocortin-4 receptor agonist causes weight loss, reduces insulin resistance, and improves cardiovascular function in diet-induced obese rhesus macaques. Diabetes. 2013;**62**:490-497. DOI: 10.2337/db12-0598

[38] Clemmensen C, Finan B, Fischer K, Tom RZ, Legutko B, Sehrer L, et al. Dual melanocortin-4 receptor and GLP-1 receptor agonism amplifies metabolic benefits in diet-induced obese mice. EMBO Molecular Medicine. 2015;**7**:288-298. DOI: 10.15252/emmm.201404508

[39] Altan VM, Arioglu E, Guner S, Ozcelikay AT. The influence of diabetes on cardiac β-adrenoceptor subtypes. Heart Failure Reviews. 2007;**12**:58-65. DOI: 10.1007/s10741-007-9005-6

[40] Chavali V, Tyagi SC, Mishra PK. Predictors and prevention of diabetic cardiomyopathy. Diabetes, Metabolic Syndrome and Obesity. 2013;**6**:151-160. DOI: 10.2147/DMSO.S3096

[41] Dincer UD, Bidasee KR, Guner S, Tay A, Ozçelikay AT, Altan VM. The effect of diabetes on expression of β1-, β2- and β3-adrenoreceptors in rat hearts. Diabetes. 2001;**50**:455-461. DOI: 10.2337/diabetes.50.2.455

[42] Derkach KV, Shpakov AO, Moyseyuk IV, Chistyakova OV. The functional activity of adenylyl cyclase signaling system in the brain, myocardium, and testes of rats with 8- and 18-month neonatal diabetes. Doklady Biochemistry and Biophysics. 2013;**448**:43-45. DOI: 10.1134/S1607672913010122

[43] Shpakov AO, Derkach KV. The functional state of hormone-sensitive adenylyl cyclase signaling system in diabetes mellitus. Journal of Signal Transduction. 2013;**2013**:594213. DOI: 10.1155/2013/594213

[44] Shpakov AO, Derkach KV, Chistyakova OV, Moyseyk IV, Bondareva VM. The effect of long-term diabetes mellitus induced by treatment with streptozotocin in 6-week-old rats on functional activity of the adenylyl cyclase system. Cell and Tissue Biology. 2014;**8**:68-79. DOI: 10.1134/S1990519X14010118

[45] Matsuda N, Hattori Y, Gando S, Akaishi Y, Kemmotsu O, Kanno M. Diabetes-induced down-regulation of β1-AR mRNA expression in rat heart. Biochemical Pharmacology. 1999;**58**:881-885. DOI: 10.1016/S0006-2952(99)00164-1

[46] Rozec B, Gauthier C. β_3-Adrenoceptors in the cardiovascular system: Putative roles in human pathologies. Pharmacology & Therapeutics. 2006;**111**:652-673. DOI: 10.1016/j.pharmthera.2005.12.002

[47] Sakane N, Yoshida T, Yoshioka K, Nakamura Y, Umekawa T, Kogure A, et al. $Trp^{64}Arg$ mutation of β_3-adrenoceptor gene is associated with diabetic nephropathy in type II diabetes mellitus. Diabetologia. 1998;**41**:1533-1534. PMID: 9867224

[48] Communal C, Singh K, Sawyer DB, Colucci WS. Opposing effects of β1- and β2-adrenergic receptors on cardiac myocyte apoptosis. Role of pertussis-toxin sensitive G protein. Circulation. 1999;**100**:2210-2212. DOI: 10.1161/01.CIR.100.22.2210

[49] Shpakov AO, Kuznetsova LA, Plesneva SA, Pertseva MN. Molecular mechanisms of modified sensitivity of the adenylate cyclase signaling system to biogenic amines during streptozotocin-induced diabetes. Bulletin of Experimental Biology and Medicine. 2005;**140**:304-308. DOI: 10.1007/s10517-005-0474-0

[50] Shpakov AO, Derkach KV, Moyseyuk IV, Chistyakova OV. Alterations of hormone-sensitive adenylyl cyclase system in the tissues of rats with long-term streptozotocin diabetes and the influence of intranasal insulin. Dataset Papers in Pharmacology. 2013;**2013**:698435. DOI: 10.7167/2013/698435

[51] Banyasz T, Kalapos I, Kelemen SZ, Kovacs T. Changes in cardiac contractility in IDDM and NIDDM diabetic rats. General Physiology and Biophysics. 1996;**15**:357-369. PMID: 9228518

[52] Huisamen B, Marais E, Genade S, Lochner A. Serial changes in the myocardial beta-adrenergic signalling system in two models of non-insulin dependent diabetes mellitus. Molecular and Cellular Biochemistry. 2001;**219**:73-82. PMID: 11354257

[53] Shpakov AO, Kuznetsova LA, Plesneva SA, Pertseva MN. Disturbance of transduction of adenylyl cyclase-inhibiting hormonal signaling in the myocardium and brain of rats with experimental type 2 diabetes. Cell and Tissue Biology. 2007;**1**:343-351. DOI: 10.1134/S1990519X07040062

[54] Mishra PK, Awe O, Metreveli N, Qipshidze N, Joshua IG, Tyagi SC. Exercise mitigates homocysteine-β_2-adrenergic receptor interactions to ameliorate contractile dysfunction in diabetes. International Journal of Physiology, Pathophysiology and Pharmacology. 2011;**3**:97-106. PMID: 21760968

[55] Sundaresan PR, Sharma VK, Gingold SI, Banerjee SP. Decreased β-adrenergic receptors in rat heart in streptozotocin-induced diabetes: Role of thyroid hormones. Endocrinology. 1984;**114**:1358-1363. DOI: 10.1210/endo-114-4-1358

[56] Derkach KV, Kuznetsova LA, Chistyakova OV, Ignatieva PA, Shpakov AO. The effect of four-week levothyroxine treatment on hormonal regulation of adenylyl cyclase in the brain and peripheral tissues of obese rats. Biochemistry (Moscow) Suppl Ser A: Membrane Cell Biol. 2015;**9**:236-245. DOI: 10.1134/S1990747815040030

Chapter 13

Systems Evolutionary Biology of Waddington's Canalization and Genetic Assimilation

Alexander V. Spirov, Marat A. Sabirov and
David M. Holloway

Additional information is available at the end of the chapter

http://dx.doi.org/10.5772/intechopen.73662

Abstract

In recent years, there has been growing interest in computer modeling of the evolution of gene and cell regulatory networks, in general, and in computational studies of the classic ideas of Baldwin, Schmalhausen, Waddington, and followers, in particular. Two related aspects of Waddington's evolutionary theories are the concepts of canalization and of genetic assimilation. Canalization is associated with the robust development of an individual to diverse perturbations and noise, though, when fluctuations in developmental factors exceed a particular limit, the normal developmental trajectory can be "thrown out" of the robust canal, resulting in an altered phenotype. If selective pressure favors the new phenotype, an initial individual loss of canalization can lead to phenotypic changes in the population (with canalization then becoming established for the new phenotype). Genetic assimilation is the subsequent genetic fixing of the new trait in the population. Recent experimental and theoretical works have established a quantitative basis for these classic concepts of Waddington; this chapter will review these new developments in systems evolutionary biology.

Keywords: canalization, genetic assimilation, gene networks, computer modeling, systems evolutionary biology

1. Canalization and genetic assimilation

Computational studies of the classic concepts of Ivan Schmalhausen [1], Conrad Waddington [2], and their contemporaries (Rendel [3]) have become a major area in evolutionary theory in recent years. A number of these concepts from the 1950s have had a major impact on the evolutionary theory of development, and computation allows for quantitative testing and characterization of the ideas. Here, we review recent work on canalization, whereby species show low phenotypic variation, despite ample genetic and environmental variation (also termed "robustness"), and on genetic assimilation, in which a phenotypic change induced by an environmental perturbation becomes stabilized in the genotype.

Canalization captures the observation that most developmental phenotypes display a certain degree of stability, despite environmental or genetic perturbations (**Figure 1**). However, for new phenotypes to arise, there must be a limit to this robustness, such that a large enough perturbation will knock the developmental trajectory out of the robust canal, resulting in a new phenotype. If this new phenotype represents higher fitness, it can be reinforced by genetic assimilation (**Figure 2**).

Waddington did a number of perturbation experiments in *Drosophila* (fruit fly) development to characterize such canalization (robustness) and show its underlying genetic basis. In those times Waddington preferred to use simple perturbations of environmental parameters and conditions, such as exposing flies to diethyl ether [5], high sodium chloride concentrations [6], or heat shock (40°C) [7]. Later, other authors have used mutations in key developmental genes as the perturbations [8, 9]. More recent approaches include genetically engineered organisms with loss-of-function [10] or gain-of-function [11, 12] mutations and varying dosages of small interfering RNAs (siRNAs) to quantitatively deplete targeted gene products [13]. Perturbation experiments remain the main approach to study the mechanisms of robustness. Whatever the technique, relative robustness is calculated as the change in variation of one or more specific traits when the experimental perturbation is applied.

In parallel with the new experimental approaches for perturbation and observations from field biology, a large branch of systems biology is now concerned with computer modeling of

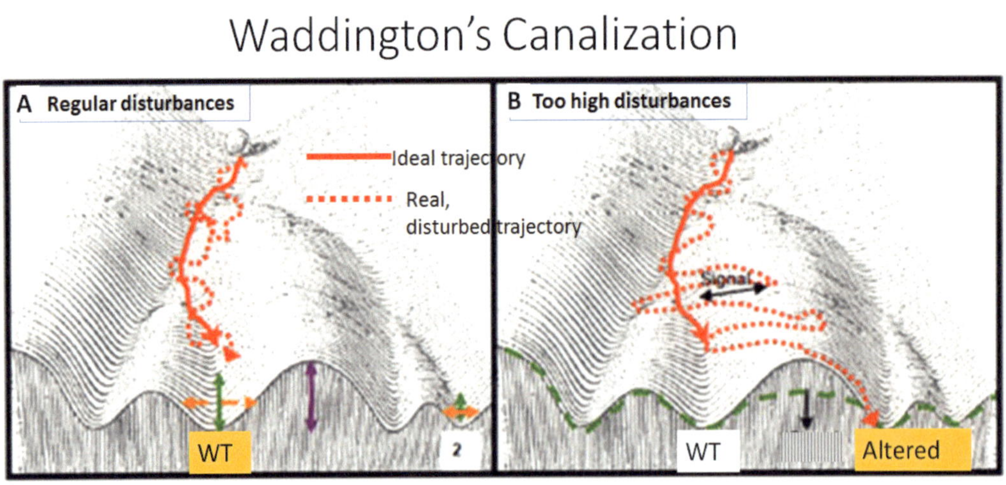

Figure 1. (A–B) Waddington's canalization and epigenetic landscape: Diverse and inevitable environmental disturbances and internal developmental noise systematically disturb developmental trajectory on the epigenetic landscape. However, the developmental process usually returns to the basin of normal development (creod), that is, the development is canalized and the canal walls keep the process in the basin prescribed by the genetic program (after http://www.gen.cam.ac.uk/research-groups/martinez-arias).

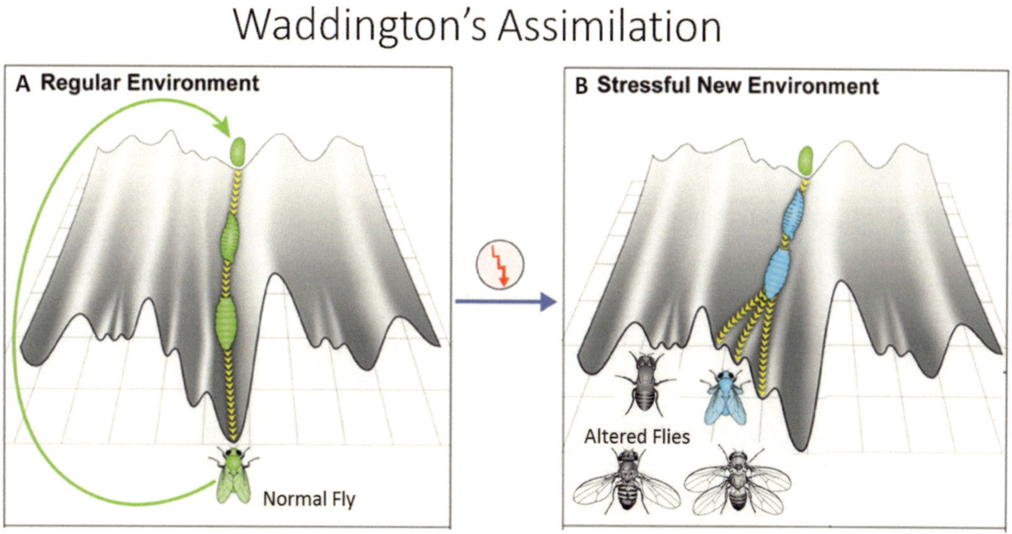

Figure 2. (A–B) Waddington's genetic assimilation: The environmental stress causes a series of the Drosophila's divergent phenotypes. The untypical, high environmental disturbances deform, change the epigenetic landscape. By doing so, it causes the appearance of new phenotypes in the population under stress. If some of the phenotypes are beneficial, it can be stabilized in the genotype by further selection (after [4]).

evolution, in order to test and verify hypotheses for evolutionary mechanisms (e.g., especially in studying evolution over numerous generations and with strictly controlled evolutionary rules). This review focuses on such work which has tested and extended the classic concepts of canalization and genetic assimilation.

2. Simulation of gene network evolution

Andreas Wagner laid out an approach to computational modeling of gene network evolution in two pioneering publications [14, 15]. In his models, an evolving population comprised individuals, each characterized by a genotype and a phenotype. Individuals have a developmental phase of their lifetimes, in which an initial phenotype develops to reach a new stable phenotype. Development is specified by the genotype, which is modeled as a gene regulatory network (GRN). Once an individual has reached its stable adult phenotype, it can reproduce to create the next generation. Reproduction occurs sexually, implying a random mixing of parental genomes in the offspring's genome. Mutations can also occur, modifying gene-gene interactions which may then disrupt the viability of the individual. Stabilization of the adult phenotype is one of the core hypotheses of the Wagner model and its subsequent variants and further developments. Some models have been proposed which constrain selection to be for a particular predetermined phenotype, rather than any stable phenotype. Or fitness functions have been studied in which fitness decreases as the Hamming distance increases between the individual's phenotype and the predetermined one [14–20]. The ways to extend and develop further the approach were reviewed recently [21].

3. How canalization evolves

Computational modeling has shown a number of different aspects of how canalization operates in evolution [21].

3.1. Phenotypic robustness to diverse perturbations

Computation allows for the separate consideration of a number of types of genetic perturbations (such as point mutations, small deletions/insertions, crossover, or gene duplications) and non-genetic perturbations (such as fluctuations in a gene circuit's cellular or nuclear microenvironment [22–24] or changes in an organism's macroenvironment).

Three major reasons for phenotypic robustness to a particular type of natural perturbation have been proposed [25, 26]. First, robustness to a perturbation might have evolved as an adaptation to reduce phenotypic variation in response to a specific perturbation. Second, robustness to the specific perturbation could be a congruent byproduct of evolved robustness to a different perturbation. Lastly, we can hypothesize that robustness is an intrinsic property of biological systems selected for their primary functions. Computational simulations of GRNs suggest both that intrinsic robustness could be widespread and that natural selection can increase robustness under diverse and reasonable sets of parameter values and assumptions [15, 16, 27–31]. Whichever of these options applies, robustness to mutation results in the accumulation of phenotypically cryptic genetic variation (CGV), that is, it allows for changes in genotype which do not affect phenotype (until a strong enough perturbation is made). Partial robustness can lead to preadaptation, and thereby might contribute to evolvability [25] (i.e., accumulated genetic variation may allow for rapid evolution under new selective pressures).

A number of studies have shown that genetic canalization would evolve under stabilizing selection [16, 32–35].

3.2. More realistic models of GRN evolution

Discrete (Boolean, "on/off") GRN models are simplified representations of gene interactions but allow for rapid analysis of some aspects of network evolution. Earlier evolutionary studies were generally with discrete models; more recent developments include continuous treatments of gene states, which are more biologically realistic but more computationally intensive to solve.

Draghi and Whitlock [36] developed a GRN model with continuous gene expression levels, affected by environmental cues, forming the phenotype. They showed by computational experiments that the evolution of phenotypic plasticity can produce populations with larger mutational variance and larger standing genetic variance. They found also that plastic populations do not respond much more quickly to selection pressure than do populations that are more static. Furthermore, if the optimal phenotypes of two traits vary together, then larger mutational and genetic correlations were observed. According to their findings, the quantitative genetic descriptions of traits created by explicit developmental network models are evolutionarily labile, with genetic correlations that change rapidly with shifts in the selection regime [36].

Iwasaki, Tsuda, and Kawata developed an individual-based approach to the GRN modeling [37, 38]. The GRN of each individual had both phenotypic and regulatory genes, each gene was composed of a cis-regulatory region and a coding region, and a cis-regulatory region was composed of cis-sites for specific transcription factors. They showed by the approach that simple GRNs tend to evolve under conditions where genetic canalization is expected, while more complex GRNs tend to evolve in conditions favoring decanalization. Iwasaki and co-authors study showed that complex GRNs display a high mutational robustness (i.e., mutations against core genes have only a small phenotypic effect) and evolvability (i.e., a larger mutational target size and mutation are likely to change the phenotype). In contrast, simple GRNs have mutational robustness only because of their small mutational target size. Iwasaki and co-authors found that the level of CGVs in a population was mainly determined by the order (weighted size) of GRNs and concluded that the outgrowth of GRNs and adaptation to new environments are mutually facilitating, resulting in sustainable evolvability [37].

3.2.1. *Evolution of genotype-phenotype mapping*

Crombach and Hogeweg [34] did a computational study with an individual-oriented model with population on a lattice subjected to an environment that changes over time. They showed that long-term evolution of complex GRNs in a changing environment can increase the generation of beneficial mutations. The population evolves toward genotype-phenotype mappings that allow for an orchestrated network-wide change in the gene expression pattern, requiring only a few specific gene *indels* (small insertions and deletions), and the genes involved are hubs of the networks or directly influencing the hubs. In addition, the GRNs maintain their mutational robustness throughout the evolutionary trajectory: evolution in a changing environment leads to a network that is sensitive to a small class of beneficial mutations, while the majority of mutations remain neutral—an example of the evolution of evolvability. These evolutionary dynamics showed a number of similarities with experimental studies in yeast (*S. cerevisiae*) [39, 40] and *E. coli*. [41].

4. Genetic assimilation

The understanding of genetic assimilation, similarly, has been extended through computational investigations in recent years.

4.1. Waddington's canalization and genetic assimilation

A heat shock perturbation done by Waddington in 1953 is used as a classic example of genetic assimilation: cross-veinless flies resulted from an initial heat shock and, selected over multiple generations, eventually produced the phenotype without the perturbation [7]. Alternatively, the cross-veinless phenotype could be due to contributions from multiple genes, which could have a lower heat shock threshold. This Falconer & Mackay threshold model [42] predicts that if selection for the assimilating cross-veinless phenotype was relaxed, genetic assimilation would not occur. Masel tested this prediction with a Wagner-type GRN model [43]. Her results indicated that genetic assimilation can occur in the absence of selection for the trait, supporting Waddington's mechanism.

4.2. Phenotypic plasticity and CGV

Closer consideration of genetic assimilation shows that it must involve some degree of phenotypic plasticity—the capacity for a genotype to produce multiple phenotypes in response to non-genetic perturbations. This plasticity may also be considered from the stand-point of release of CGV (which Masel and Trotter [44] defined as standing genetic variation that does not ordinarily contribute to the phenotype), which increases during neutral drift but may only become visible with a large enough perturbation. CGV accumulation can cause a diversification of genetic backgrounds on which new mutations may arise. It is biologically reasonable to expect that the effect of any new mutation should be background dependent: a given mutation would have a given effect at a particular background and a different or no effects at other backgrounds. The diversity of genetic backgrounds would give the population access to more novel phenotypes than if it were isogenic [45–47], as reviewed in [48]. As Siegal and Leu [48] summarized, this conceptual argument for evolvability correlating positively with mutational robustness has been formalized in mathematical models of so-called neutral networks in genotype space (more recently termed "genotype networks") and has some empirical support [49–51].

Iwasaki with co-authors [38] focused on GRN as an important mechanism for producing CGV and examined how interactions between GRNs and the environment influence the number of CGVs by using individual-based simulations. The authors conclude that interactions with variable environments may promote the accumulation of CGVs by facilitating the evolution of larger GRNs. In turn, the expansion of GRNs could facilitate evolutionary adaptation to novel environments and niche construction [38].

4.3. Phenotypic plasticity and genetic assimilation

Lande defines genetic assimilation in an altered environment as the reduction in plasticity and its replacement by genetic evolution, while maintaining the phenotype initially produced by plasticity in the altered environment [52]. According to Lande, reduction in plasticity during genetic assimilation is often attributed to fitness costs of plasticity.

Lande quantifies the relation between phenotypic plasticity dynamics and genetic assimilation, wherein plasticity must increase to allow evolution to a perturbed environment but then be reduced to maintain the new optimum. During the first generation in the novel environment, the average fitness substantially drops and the average phenotype jumps toward the new optimum by expression of partially adaptive plasticity. Then, transient evolution of increased plasticity accelerates phenotypic adaptation and allows the average phenotype to come toward the new optimum. Then, the novel phenotype undergoes a slow process of genetic assimilation, with reduction in plasticity [52].

Temporary perturbations that reduce robustness could turn unitary phenotypes into plastic ones. It then gives natural selection a substrate on which to select a particular novel phenotype (i.e., genetic assimilation) [48, 53–55]. High levels of phenotypic variation could increase the chance of population survival in new hostile environments, which in turn would give time for the population to accumulate adaptive mutations [48, 56, 57]. As Siegal, Leu concluded, the connection between robustness and plasticity could be especially important to evolution [48].

Janna Fierst asked herself to what degree can a history of phenotypic plasticity affect the rate of adaptation to a new environment, that is, is plasticity merely a condition for genetic assimilation, or do environmental fluctuations cause phenotypic plasticity, generating genotypic evolvability? She showed that a history of phenotypic plasticity may determine the evolution of genetic architecture and shorten the waiting time for the generation of phenotypic variance from new mutations and recombination. Hence, rather than acting as a short-term alternative, phenotypic plasticity may facilitate future adaptation and genetic evolution [58].

4.3.1. Phenotypic plasticity and evolvability

Non-genetic perturbations, such as environmental change or developmental noise, can induce novel phenotypes. If an induced phenotype appears recurrently and confers a fitness advantage, selection may promote its genetic stabilization. Non-genetic perturbations can thus initiate evolutionary innovation. CGV may play an important role in this process [20]. Populations under stabilizing selection on a phenotype that is robust to mutations can accumulate such variation. After non-genetic perturbations, this variation can produce new phenotypes. Espinosa-Soto with co-authors find that phenotypic robustness promotes phenotypic variability in response to non-genetic perturbations but not in response to mutation. It suggests that non-genetic perturbations may initiate innovation more frequently in mutationally robust gene expression traits [20].

Phenotypic plasticity can facilitate the origin of genotypes that produces a new phenotype in response to non-genetic perturbations. Espinosa-Soto with co-authors find that phenotypic plasticity frequently facilitates the evolution of novel beneficial gene activity patterns in gene regulatory circuits [59]. The fundamental reason is that genotypes that produce occasionally a beneficial phenotype (and thus have a selective advantage) give more easily rise to genotypes where that same phenotype is more strongly genetically determined [59].

The characterization of plasticity, robustness, and evolvability can be studied in terms of phenotypic fluctuations. By numerically evolving GRNs, the proportionality between the phenotypic variances of epigenetic and genetic origins is confirmed by Kaneko [60]. The relationship suggests a link between robustness to noise and to mutation. The proportionality between the variances is demonstrated to also hold over expressions of different genes (phenotypic traits) when the system acquires robustness through the evolution. It was found by Kaneko that both the population's adaptability to a new environment and the population's robustness becomes compatible when a certain degree of phenotypic fluctuations is produced by the developmental variability and noise [60]. The Kaneko's conclusion is that the highest adaptability is achieved at near-the-threshold noise level at which the gene expression dynamics are near the critical point to lose the robust evolutionary process.

5. Canalization and assimilation in population biology

Current advances in evolutionary systems biology were caused not only by working out of new computational approaches but also by new biological observations performed to verify the computational conclusions.

5.1. CGV in natural populations

Biological systems are highly robust to perturbation by mutations, recombination, and the environmental stress. Robustness to mutation and recombination permits genetic variation to accumulate as hidden genetic diversity (or CGV). CGV might be revealed in response to stress, and "the amount of heritable phenotypic variation available can be correlated to the degree of stress and hence to the novelty of the environment…" [44].

The CGV role in genetic assimilation was extensively studied by computational evolutionary tests (as overviewed in Section 4.2). They are considered to contribute to evolutionary responses to environmental changes by generating phenotypic diversity [61–63]. Furthermore, the CGV's ability to accumulate and release multiple mutations in populations supports some researchers' considerations that CGVs also promote the acquisition of new traits [38, 64, 65].

Some experiments support these considerations. For example, as it was shown by Suzuki and Nijhout [55], a mutation in the larval hormone-regulatory pathway in *Manduca sexta* moth enables heat stress to reveal a hidden larval coloration. The *black* mutant strain of the moth, which was originally green, demonstrated variations in thermosensitivity: heat shocks during the sensitive period generated larvae with colors that ranged from normal black to nearly normal green [55]. Suzuki and Nijhout also successfully established two lines by artificial selection: one selected for increased greenness upon heat treatment (sensitive line) and the other for decreased color change upon heat treatment (insensitive line). Hence, CGVs really could contribute to phenotypic evolution.

5.2. Phenotypic capacitors

Phenotypic capacitor "is a biological switch capable of revealing previously cryptic heritable variation" [25]. This is an analogy with an electric capacitor, which is capable to store and release an electric charge. Many of the capacitors are proteins whose function contributes to robustness and, therefore, whose damage or modification reveals phenotypic variation [66].

In a complex GRN, there are many gene products which could appear as "phenotypic" capacitors, such that their removal increased phenotypic variability. An extensively studied example is the molecular chaperone Hsp90, but GRN dynamics indicated there should be many more. Experiments in yeast indicated more than 300 gene products whose removal increased variation [10].

In Drosophila, other molecular chaperones—Hsp22, Hsp67, and Hsp70—were also observed to affect either within-individual variation (measured by asymmetry of bilateral traits) or among-individual variation in morphology [67]. In eukaryotes, Hsp90 impairment has been found to reveal CGV in organisms ranging from yeast to flies to vertebrates to plants [68, 69].

Masel and Siegal [25] considers three approaches based on the use of phenotypic capacitors to study robustness. The first approach is a genome-wide screening for genetic perturbations affecting the variance of a given trait. The trait can be morphological [10, 70, 71], or physiological [72] or can be measured as RNA and protein concentrations [73–75]. Good examples of the approach include the studies of cellular morphology in *S. cerevisiae* mutant strains [10, 76] and the genome-wide analysis of more complex and quantitative traits in both *S. cerevisiae* [70, 73] and *A. thaliana* [71, 74].

The second approach is based on usage of a well-characterized model developmental system under the impact of perturbations. Good example of the approach is the consideration of the developmental lineage of the cells comprising the vulva of nematode species of the genus *Caenorhabditis* [77, 78]. Perturbation of *C. elegans* vulva development by mutation or environmental variation revealed changes in the underlying signaling pathways [77, 78]. Robustness of the vulval developmental system to environmental perturbations results through an integration of multiple buffering capacities at the molecular and cellular level [77, 78].

The third complementary approach is to focus on a single well-characterized perturbation and the variety of developmental systems that it affects. Examples include perturbation of translation termination by the yeast prion [*PSI*+] [79–81] and the heat shock protein, Hsp90, which affects a stunning variety of developmental processes [70, 82].

Namely the computational evolutionary experiments with the GRN models revealed possible existence and evolutionary significance of the phenotypic capacitors and brought intent attention to its experimental study.

5.3. Phenotypic plasticity and genetic assimilation in biology

Baldwin [56], Simpson [83], and Waddington [84, 85] proposed that phenotypic plasticity may benefit populations in new environments. In accordance with Waddington's pioneer considerations, artificial selection can turn an alternative phenotype into a native one [5, 7]. More recently, other researchers have confirmed his observation for diverse traits and different species [55, 70, 86].

Many empirical studies of wild populations support the hypothesis that an ancestral alternative phenotype could have facilitated the evolution of novel, adaptive traits [16, 28, 29, 72, 87–93]. For example, severe environments enhance phenotypic differences among fruit fly strains [94], and a temperature rise caused by a lack of shade increases the frequency of abnormal morphologies in fruit flies [95].

The phenotypes where plasticity may have facilitated adaptation are very diverse. They include gill surface area in cichlid fishes [96], pigmentation patterns in the crustacean *Daphnia melanica* [97], and head size in the snake *Notechis scutatus* [98]. Despite an abundance of candidate examples, plasticity's importance for adaptive evolution is not universally accepted, and we still do not know whether existing observations are rare oddities or hint at general principles of evolution [99–105].

5.3.1. Natural populations in changing environments

During millions of years of existence, species repeatedly encounter extreme changes in average environment, and the capacity to accelerate phenotypic adaptation by transient evolution of plasticity may be crucial for long-term persistence. Sudden environmental change often occurs at the start of natural biological invasions and colonizations (reviewed in [106]).

The success of natural invasions, and artificial introductions for biocontrol, may depend on the evolution of increased plasticity during adaptation to novel environments outside the native range of a species [107–109]. Genetic variance in plasticity within and/or among populations has commonly been observed [110, 111], and species invading novel or extreme environments often display increased plasticity compared to populations from the native range [96, 112–117]. Populations of invasive species outside their native range usually maintain substantial genetic variance [118–120].

Experiments on newly established small populations show that intense artificial selection can rapidly create large phenotypic changes, often altering the mean phenotype by several standard deviations within a few dozen generations [121–123]. For extremely large populations undergoing sudden environmental change *in situ*, sustained intense directional selection can cause adaptation by a rare allele of major effect [124, 125].

Many empirical studies suggest that invasive species tend to have an evolutionary history of environmental disturbance [53]. Ecological disturbances constitute fluctuating selection pressures over evolutionary time, and evolutionary genetic theory predicts that patterns of fluctuating selection can cause genetic architectures to take different paths (e.g. [126]).

When environmental changes happen infrequently, populations maximize fitness by producing a single phenotype [58]. When the environment changes more frequently, organisms that can evolve more rapidly are favored by selection. As Janna Fierst concluded, "when environmental fluctuations are rapid, fitness is maximized by genetic architectures that produce a broad, generalist phenotype or short-term phenotypic plasticity" [58].

Increasing amounts of evidence suggest that traits induced by non-genetic factors are important for innovation in nature [98, 127–129]. For example, taxa with genetically determined dextral or sinistral morphologies are frequently derived from taxa in which the direction of the asymmetry is not genetically fixed but where it is a plastic response [128, 130]. Transitions like these imply genetic assimilation of a direction of asymmetry. This was observed for multiple traits, such as the side on which the eye occurs in flat fishes (*Pleuronectiformes*) and the side of the larger first claw in decapods (*Thalassinidea*) [128].

More generally, good candidates for genetic assimilation are the traits where fixed differences among closely related species are mirrored by plastic variation within populations. Amphibian traits, such as gut morphology [129], limb length, and snout length [130], are illustrative examples.

We can conclude that the observations on the environmental dependence of phenotypic and genetic variances evidences accelerated phenotypic adaptation after an extraordinary environmental change [52].

Acknowledgements

The results of Section 5 were obtained within the state assignment of FASO of Russia (theme No. 01201351572). The results of all other sections were obtained within the RSF grant (project No. 17-18-01536).

Author details

Alexander V. Spirov[1]*, Marat A. Sabirov[1] and David M. Holloway[2,3]

*Address all correspondence to: spirov@iephb.nw.ru

1 Lab Modeling of Evolution, I.M. Sechenov Institute of Evolutionary Physiology and Biochemistry, Russian Academy of Sciences, St. Petersburg, Russia

2 Mathematics Department, British Columbia Institute of Technology, Burnaby, Canada

3 Biology Department, University of Victoria, Victoria, Canada

References

[1] Schmalhausen II. Factors of Evolution: The Theory of Stabilizing Selection. Chicago: Univ. of Chicago Press; 1986

[2] Waddington CH. The Strategy of the Genes. A Discussion of some Aspects of Theoretical Biology. London: Allen and Unwin; 1957

[3] Rendel JM. Canalization of the acute phenotype of Drosophila. Evolution. 1959;**13**:425-439

[4] Soen Y. Environmental disruption of host-microbe co-adaptation as a potential driving force in evolution. Frontiers in Genetics. 2014;**5**(168)

[5] Waddington CH. Genetic assimilation of the bithorax phenotype. Evolution. 1956;**10**(1):13

[6] Waddington CH. Canalization of development and genetic assimilation of acquired characters. Nature. 1959;**183**(4676):1654-1655

[7] Waddington CH. Genetic assimilation of an acquired character. Evolution. 1953;**7**:118-126

[8] Rutherford SL, Lindquist S. Hsp90 as a capacitor for morphological evolution. Nature. 1998;**396**:336-342

[9] Scharloo W. Canalization: Genetic and developmental aspects. Annual Review of Ecology and Systematics. 1991;**22**:65-93

[10] Levy SF, Siegal ML. Network hubs buffer environmental variation in *Saccharomyces cerevisiae*. PLoS Biology. 2008;**6**(11):e264

[11] Dworkin I, Palsson A, Birdsall K, Gibson G. Evidence that Egfr contributes to cryptic genetic variation for photoreceptor determination in natural populations of *Drosophila melanogaster*. Current Biology. 2003;**13**:1888-1893

[12] Polaczyk PJ, Gasperini R, Gibson G. Naturally occurring genetic variation affects Drosophila photoreceptor determination. Development Genes and Evolution. 1998;**207**(7): 462-470

[13] Baggs JE, Price TS, DiTacchio L, Panda S, FitzGerald GA, Hogenesch JB. Network features of the mammalian circadian clock. PLoS Biology. 2009;**7**(3):e1000052

[14] Wagner A. Evolution of gene networks by gene duplications: A mathematical model and its implications on genome organization. Proceedings of the National Academy of Sciences of the United States of America. 1994;**91**:4387-4391

[15] Wagner A. Does evolutionary plasticity evolve? Evolution. 1996;**50**:1008-1023

[16] Siegal ML, Bergman A. Waddington's canalization revisited: Developmental stability and evolution. Proceedings of the National Academy of Sciences of the United States of America. 2002;**99**:10528-10532

[17] Kimbrell T, Holt RD. Canalization breakdown and evolution in a source-sink system. The American Naturalist. 2007;**169**:370-382

[18] MacCarthy T, Bergman A. Co-evolution of epistasis and recombination favors asexual reproduction. PNAS. 2007;**104**:12801-12806

[19] Lohaus R, Burch CL, Azevedo RBR. Genetic architecture and the evolution of sex. The Journal of Heredity. 2010;**101**(Suppl. 1):S142-S157

[20] Espinosa-Soto C, Martin OC, Wagner A. Phenotypic plasticity can increase phenotypic variability after non-genetic perturbations in gene regulatory circuits. Journal of Evolutionary Biology. 2011;**24**:1284-1297

[21] Spirov A, Holloway D. Using EA to study the evolution of GRNs controlling biological development. In: Noman N, Iba H, editors. Evolutionary Algorithms in Gene Regulatory Network Research. Wiley Interscience; 2015. pp. 240-268

[22] McAdams HH, Arkin A. Stochastic mechanisms in gene expression. Proceedings of the National Academy of Sciences of the United States of America. 1997;**94**:814-819

[23] Elowitz MB, Levine AJ, Siggia ED, Swain PS. Stochastic gene expression in a single cell. Science. 2002;**297**(5584):1183-1186

[24] Raj A, Rifkin SA, Andersen E, van Oudenaarden A. Variability in gene expression underlies incomplete penetrance. Nature 2010;**463**:913-918

[25] Masel J, Siegal ML. Robustness: Mechanisms and consequences. Trends in Genetics. 2009;**25**:395-403

[26] de Visser J, et al. Perspective: Evolution and detection of genetic robustness. Evolution 2003;**57:**1959-1972

[27] Kim KJ, Fernandes VM. Effects of Ploidy and recombination on evolution of robustness in a model of the segment polarity network. PLoS Computational Biology. 2009;**5**: e1000296

[28] Proulx SR, Phillips PC. The opportunity for canalization and the evolution of genetic networks. The American Naturalist. 2005;**165**:147-162

[29] Soyer OS et al. Simulating the evolution of signal transduction pathways. Journal of Theoretical Biology. 2006;**241**:223-232

[30] van Nimwegen E, et al. Neutral evolution of mutational robustness. Proceedings of the National Academy of Sciences of the United States of America 1999;**96:**9716-9720

[31] Wagner A. Circuit topology and the evolution of robustness in two-gene circadian oscillators. Proceedings of the National Academy of Sciences of the United States of America. 2005;**102**:11775-11780

[32] Kawecki TJ. The evolution of genetic canalization under fluctuating selection. Evolution. 2000;**54**:1-12

[33] Martınez-Antonio A, Janga SC, Thieffry D. Functional organisation of *Escherichia coli* transcriptional regulatory network. Journal of Molecular Biology. 2008;**381**:238-247

[34] Crombach A, Hogeweg P. Evolution of evolvability in gene regulatory networks. PLoS Computational Biology. 2008;**4**:e1000112

[35] Kashtan N, Noor E, Alon U. Varying environments can speed up evolution. Proceedings of the National Academy of Sciences of the United States of America. 2007;**104**:13711-13716

[36] Draghi JA, Whitlock MC. Phenotypic plasticity facilitates mutational variance, genetic variance, and evolvability along the major axis of environmental variation. Evolution. 2012;**66**(9):2891-2902

[37] Tsuda ME, Kawata M. Evolution of gene regulatory networks by fluctuating selection and intrinsic constraints. PLoS Computational Biology. 2010;**6**(8):e1000873

[38] Iwasaki WM, Tsuda ME, Kawata M. Genetic and environmental factors affecting cryptic variations in gene regulatory networks. BMC Evolutionary Biology. 2013;**13**:91

[39] Ferea T, Botstein D, Brown P, Rosenzweig R. Systematic changes in gene expression patterns following adaptive evolution in yeast. Proceedings of the National Academy of Sciences of the United States of America. 1999;**96**:9721-9726

[40] Dunham M, Badrane H, Ferea T, Adams J, Brown P, et al. Characteristic genome rearrangements in experimental evolution of *Saccharomyces cerevisiae*. Proceedings of the National Academy of Sciences of the United States of America. 2002;**99**:16144-16149

[41] Philippe N, Crozat E, Lenski RE, Schneider D. Evolution of global regulatory networks during a long–term experiment with *Escherichia coli*. BioEssays. 2007;**29**:846-860

[42] Falconer DS, Mackay TFC. Introduction to Quantitative Genetics. Essex: Longman; 1996. pp. 309-310

[43] Masel J. Genetic assimilation can occur in the absence of selection for the assimilating phenotype, suggesting a role for the canalization heuristic. Journal of Evolutionary Biology. 2004;**17**(5):1106-1110

[44] Masel J, Trotter MV. Robustness and Evolvability. Trends in Genetics. 2010;**26**(9):406-414

[45] Wagner A. Robustness and Evolvability in Living Systems. Princeton Univ. Press: Princeton; 2007. p. 367

[46] Wagner A. The molecular origins of evolutionary innovations. Trends in Genetics. 2011; **27**:397-410

[47] Wagner A. The role of robustness in phenotypic adaptation and innovation. Proceedings of the Biological Sciences. 2012;**279**:1249-1258

[48] Siegal ML, Leu JY. On the nature and evolutionary impact of phenotypic robustness mechanisms. Annual Review of Ecology, Evolution, and Systematics. 2014;**45**:496-517

[49] McBride RC, Ogbunugafor CB, Turner PE. Robustness promotes evolvability of thermotolerance in an RNA virus. BMC Evolutionary Biology. 2008;**8**:231

[50] Hayden EJ, Ferrada E, Wagner A. Cryptic genetic variation promotes rapid evolutionary adaptation in an RNA enzyme. Nature. 2011;**474**:92-95

[51] Lauring AS, Frydman J, Andino R. The role of mutational robustness in RNA virus evolution. Nature Reviews. Microbiology. 2013;**11**:327-336

[52] Lande R. Adaptation to an extraordinary environment by evolution of phenotypic plasticity and genetic assimilation. Journal of Evolutionary Biology. 2009;**22**:1435-1446

[53] Lee CE, Gelembiuk GW. Evolutionary origins of invasive populations. Evolutionary Applications. 2008;**1**:427-448

[54] Behera N, Nanjundiah V. Phenotypic plasticity can potentiate rapid evolutionary change. Journal of Theoretical Biology. 2004;**22**:177-184

[55] Suzuki Y, Nijhout HF. Evolution of a polyphenism by genetic accommodation. Science. 2006;**311**(5761):650-652

[56] Baldwin JM. A new factor in evolution. The American Naturalist. 1896;**30**:441-451

[57] Price TD, Qvarnstrom A, Irwin DE. The role of phenotypic plasticity in driving genetic evolution. Proceedings of the Royal Society of London B: Biological Sciences. 2003;**27**: 1433-1440

[58] Fierst JL. A history of phenotypic plasticity accelerates adaptation to a new environment. Journal of Evolutionary Biology. 2011;**24**:1992-2001

[59] Espinosa-Soto C, Martin OC, Wagner A. Phenotypic plasticity can facilitate adaptive evolution in gene regulatory circuits. BMC Evolutionary Biology. 2011;**11**(5)

[60] Kaneko K. Evolution of robustness and plasticity under environmental fluctuation: Formulation in terms of phenotypic variances. Journal of Statistical Physics. 2012;**148**(4):687-705

[61] Le Rouzic A, Carlborg O. Evolutionary potential of hidden genetic variation. Trends in Ecology & Evolution. 2008;**23**:33-37

[62] Schlichting CD. Hidden reaction norms, cryptic genetic variation, and evolvability. Annals of the New York Academy of Sciences. 2008;**1133**:187-203

[63] Wagner A. Robustness and evolvability: A paradox resolved. Proceedings of the Biological Sciences. 2008;**275**(1630):91-100

[64] West-Eberhard MJ. Developmental plasticity and the origin of species differences. Proceedings of the National Academy of Sciences of the United States of America. 2005; **102**(Suppl 1):6543-6549

[65] Pigliucci M, Murren CJ. Genetic assimilation and a possible evolutionary paradox: Can macroevolution sometimes be so fast as to pass us by? Evolution. 2003;**57**(7):1455-1464

[66] Masel J. Q&A: Evolutionary capacitance. BMC Biology. 2013;**11**:103

[67] Takahashi KH, Daborn PJ, Hoffmann AA, Takano-Shimizu T. Environmental stress-dependent effects of deletions encompassing Hsp70Ba on canalization and quantitative trait asymmetry in *Drosophila melanogaster*. PLoS One. 2011;**6**:e17295

[68] Jarosz DF, Taipale M, Lindquist S. Protein homeostasis and the phenotypic manifestation of genetic diversity: Principles and mechanisms. Annual Review of Genetics. 2010; **44**:189-216

[69] Rohner N, Jarosz DF, Kowalko JE, YoshizawaM JWR. Cryptic variation in morphological evolution: HSP90 as a capacitor for loss of eyes in cavefish. Science. 2013;**342**:1372-1375

[70] Nogami S et al. Genetic complexity and quantitative trait loci mapping of yeast morphological traits. PLoS Genetics. 2007;**3**:e31

[71] Hall MC et al. Genetics of microenvironmental canalization in *Arabidopsis thaliana*. Proceedings of the National Academy of Sciences of the United States of America. 2007;**104**: 13717-13722

[72] Cooper T et al. Effect of random and hub gene disruptions on environmental and mutational robustness in *Escherichia coli*. BMC Genomics. 2006;**7**:237

[73] Ansel J et al. Cell-to-cell stochastic variation in gene expression is a complex genetic trait. PLoS Genetics. 2008;**4**:e1000049

[74] Fu J et al. System-wide molecular evidence for phenotypic buffering in Arabidopsis. Nature Genetics. 2009;**41**:166-167

[75] Brem RB, Kruglyak L. The landscape of genetic complexity across 5,700 gene expression traits in yeast. Proceedings of the National Academy of Sciences of the United States of America. 2005;**102**:1572-1577

[76] Ohya Y et al. High-dimensional and large-scale phenotyping of yeast mutants. Proceedings of the National Academy of Sciences of the United States of America. 2005;**102**: 19015-19020

[77] Braendle C, Felix MA. Plasticity and errors of a robust developmental system in different environments. Developmental Cell. 2008;**15**:714-724

[78] Milloz J et al. Intraspecific evolution of the intercellular signaling network underlying a robust developmental system. Genes & Development. 2008;**22**:3064-3075

[79] Joseph SB, Kirkpatrick M. Effects of the [PSI+] prion on rates of adaptation in yeast. Journal of Evolutionary Biology. 2008;**21**:773-780

[80] Masel J, Bergman A. The evolution of the evolvability properties of the yeast prion [PSI +]. Evolution. 2003;**57**:1498-1512

[81] True HL, Lindquist SL. A yeast prion provides a mechanism for genetic variation and phenotypic diversity. Nature. 2000;**407**:477-483

[82] Queitsch C et al. Hsp90 as a capacitor of phenotypic variation. Nature. 2002;**417**:618-624

[83] Simpson GG. The Baldwin effect. Evolution. 1953;**7**:110-117

[84] Waddington CH. Canalization of development and the inheritance of acquired characters. Nature. 1942;**150**:563-565

[85] Waddington CH. Genetic assimilation. Advances in Genetics. 1961;**10**:257-293

[86] Wagner A. Distributed robustness versus redundancy as causes of mutational robustness. BioEssays. 2005;**27**:176-188

[87] Hopfield JJ. Kinetic proofreading: A n ew mechanism for reducing errors in biosynthetic processes requiring high specificity. Proceedings of the National Academy of Sciences of the United States of America. 1974;**71**:4135-4139

[88] Milton CC et al. Quantitative trait symmetry independent of Hsp90 buffering: Distinct modes of genetic canalization and developmental stability. Proceedings of the National Academy of Sciences of the United States of America. 2003;**100**:13396-13401

[89] Ferrell J. Q&A: Cooperativity. Journal of Biology. 2009;**8**:53

[90] Pierce SB et al. Drosophila growth and development in the absence of dMyc and dMnt. Developmental Biology. 2008;**315**:303-316

[91] O'Neill JS. Circadian clocks can take a few transcriptional knocks. The EMBO Journal. 2009;**28**:84-85

[92] Takahashi S, Pryciak PM. Membrane localization of scaffold proteins promotes graded signaling in the yeast MAP kinase cascade. Current Biology. 2008;**18**:1184-1191

[93] Kafri R et al. Transcription control reprogramming in genetic backup circuits. Nature Genetics. 2005;**37**:295-299

[94] Kondrashov AS, Houle D. Genotype-environment interactions and the estimation of the genomic mutation rate in *Drosophila melanogaster*. Proceedings of the Biological Sciences. 1994;**258**:221-227

[95] Roberts SP, Feder ME. Natural hyperthermia and expression of the heat shock protein Hsp70 affect developmental abnormalities in *Drosophila melanogaster*. Oecologia. 1999; **121**:323-329

[96] Chapman LJ, Galis F, Shinn J. Phenotypic plasticity and the possible role of genetic assimilation: Hypoxia-induced trade-offs in the morphological traits of an African cichlid. Ecology Letters. 2000;**3**(5):387-393

[97] Scoville AG, Pfrender ME. Phenotypic plasticity facilitates recurrent rapid adaptation to introduced predators. Proceedings of the National Academy of Sciences of the United States of America. 2010;**107**:4260-4263

[98] Aubret F, Shine R. Genetic assimilation and the postcolonization erosion of phenotypic plasticity in island tiger snakes. Current Biology. 2009;**19**:1932-1936

[99] Orr HA. An evolutionary dead end? Science. 1999;**285**:343-344

[100] de Jong G, Crozier RH. A flexible theory of evolution. Nature 2003;**424**:16-17

[101] Moczek AP. Developmental capacitance, genetic accommodation, and adaptive evolution. Evolution & Development. 2007;**9**:299-305

[102] Moczek AP. On the origins of novelty in development and evolution. BioEssays. 2008; **30**(5):432-447

[103] Pfennig DW, Wund MA, Snell-Rood EC, Cruickshank T, Schlichting CD, Moczek AP. Phenotypic plasticity's impacts on diversification and speciation. Trends in Ecology & Evolution. 2010;**25**(8):459-467

[104] Braendle C, Flatt T. A role for genetic accommodation in evolution? BioEssays. 2006;**28**: 868-873

[105] Hall BK. Organic selection: Proximate environmental effects on the evolution of morphology and behaviour. Biology and Philosophy. 2001;**16**:215-237

[106] Lande R. Adaptation to an extraordinary environment by evolution of phenotypic plasticity and genetic assimilation. Journal of Evolutionary Biology. 2009;**22**:1435-1446

[107] Richards CL, Bossdorf O, Muth NZ, Gurevitch J, Pigliucci M. Jack of all trades, master of some? On the role of phenotypic plasticity in plant invasions. Ecology Letters. 2006;**9**: 981-993

[108] Yoshida T, Goka K, Ishihama F, Ishihara M, Kudo S. Biological invasion as a natural experiment of the evolutionary process: Introduction to the special feature. Ecological Research. 2007;**22**:849-854

[109] Price T, Sol D. Introduction: Genetics of colonizing species. The American Naturalist. 2008;**172**:S1-S3

[110] Scheiner SM. Genetics and evolution of phenotypic plasticity. Annual Review of Ecology and Systematics. 1993;**24**:35-68

[111] Scheiner SM. Selection experiments and the study of phenotypic plasticity. Journal of Evolutionary Biology. 2002;**15**:889-898

[112] Lee CE, Remfert JL, Gelembiuk GW. Evolution of physiological tolerance and performance during freshwater invasions. Integrative and Comparative Biology. 2003;**43**:439-449

[113] Dybdhal MF, Kane SL. Adaptation vs. phenotypic plasticity in the success of a clonal invader. Ecology. 2005;**86**:1592-1601

[114] Chun YJ, Collyer ML, Moloney KA, Nason JD. Phenotypic plasticity of native vs. invasive purple loosestrife: A two-state multivariate approach. Ecology. 2007;**88**:1499-1512

[115] Cano L, Escarre J, Fleck I, Blanco-Moreno JM, Sans FX. Increased fitness and plasticity of an invasive species in its introduced range: A study using Senecio pterophorus. Journal of Ecology. 2008;**96**:468-476

[116] Lardies MA, Bozinovic F. Genetic variation for plasticity in physiological and life-history traits among populations of an invasive species, the terrestrial isopod Porcellio Laevis. Evolutionary Ecology Research. 2008;**10**:747-762

[117] Lombaert E, Malausa T, Devred R, Estoup A. Phenotypic variation in invasive and biocontrol populations of the harlequin ladybird, *Harmonia axyridis*. BioControl. 2008; **53**:89-102

[118] Sexton JP, McKay JK, Sala A. Plasticity and genetic diversity may allow saltcedar to invade cold climates in North America. Ecological Applications. 2002;**12**:1652-1660

[119] Clements DR, DiTommaso A, Jordan N, Booth BD, Cardina J, Doohan D, Mohler CL, Murphy SD, Swanton CJ. Adaptability of plants invading north American cropland. Agriculture, Ecosystems and Environment. 2004;(3):379-398

[120] Gilchrist GW, Lee CE. All stressed out and nowhere to go: Does evolvability limit adaptation in invasive species? Genetica. 2007;**129**:127-132

[121] Wright S. Evolution and the Genetics of Populations. 3. Experimental Results and Evolutionary Deductions. Chicago: University of Chicago Press; 1977

[122] Robertson A. Selection Experiments in Laboratory and Domestic Animals. Slough: Commonwealth Agricultural Bureaux; 1980

[123] Roush RT, McKenzie JA. Ecological genetics of insecticide and acaracide resistance. Annual Review of Entomology. 1987;**32**:361-380

[124] Roush RT, Tabashnik BE. Pesticide Resistance in Arthropods. London: Chapman and Hall; 1990

[125] West-Eberhard MJ. Developmental Plasticity and Evolution. Oxford: Oxford University Press; 2003

[126] Meyers LA, Ancel FD, Lachmann M. Evolution of genetic potential. PLoS Computational Biology. 2005;**1**:236-243

[127] Palmer AR. Symmetry breaking and the evolution of development. Science. 2004;**306**: 828-833

[128] Palmer AR. From symmetry to asymmetry: Phylogenetic patterns of asymmetry variation in animals and their evolutionary significance. Proceedings of the National Academy of Sciences of the United States of America. 1996;**93**:14279-14286

[129] Ledon-Rettig CC, Pfennig DW, Nascone-Yoder N. Ancestral variation and the potential for genetic accommodation in larval amphibians: Implications for the evolution of novel feeding strategies. Evolution & Development. 2008;**10**:316-325

[130] Gomez-Mestre I, Buchholz DR. Developmental plasticity mirrors differences among taxa in spadefoot toads linking plasticity and diversity. Proceedings of the National Academy of Sciences of the United States of America. 2006;**103**:19021-19026

Chapter 14

3D Structures and Molecular Evolution of Ion Channels

Denis B. Tikhonov and Boris S. Zhorov

Additional information is available at the end of the chapter

http://dx.doi.org/10.5772/intechopen.73665

Abstract

Ion channels mediate selective passive transport of ions across biomembranes. They participate in diverse physiological processes and belong to distinct protein families. Understanding specific roles of different channels in physiology, pathology, and pharmacology requires knowledge of their origin and evolution. Traditional approaches include experimental physiological studies and analysis of sequences and genomes. In the last two decades, availability of 3D structures of many ion channel proteins revolutionized ion channel studies, including their evolution. In this chapter, we consider examples of how 3D structures provided clues for understanding evolutionary aspects of multi-domain organization, domain folding, and roles of highly conserved and variable residues. Such achievements are important for addressing practical problems including drug design, channelopathies, and acquired resistance to insecticides.

Keywords: evolution physiology, ion channels, sequence alignment, folding, domains

1. Introduction

The human genome encodes hundreds of proteins that form ion channels. These proteins create transmembrane pores through which inorganic ions, mainly Na^+, K^+, Ca^{2+} or Cl^-, permeate according to their electrochemical gradient. Selectivity of the channels for particular ions, mechanisms of their activation and kinetic characteristics are greatly variable. This variability underlines involvement of ion channels in very diverse fundamental physiological processes such as generation of the membrane potential, regulation of cell electrical excitability, propagation of the action potential along neurons and muscle cells, synaptic release of neurotransmitters, generation of postsynaptic response, and calcium signaling. Ion channels are indispensable in multicellular organisms. They are also found in various bacteria and even some viruses [1]. Ion channels are targets for numerous endogenous ligands including hormones and neurotransmitters, as well as for multiple exogenous toxins and medically important and illicit drugs. Due to these roles, ion channels are key objects in pharmacology and toxicology.

This functional importance of ion channels motivates their intensive studies in academia and industry. A large body of experimental data is accumulated on selectivity and permeability characteristics of different channels, physiological mechanisms of their activation by different factors, including membrane voltage, various endogenous and exogenous ligands, metal ions, temperature, pH, and membrane tension. Selective sensitivity of ion channels to various pharmacological agents allowed to reveal their presence in various organs and tissues, cellular localization, and specific roles in physiological and pathological processes. The next big step was understanding molecular characteristics of ion channels due to methodological advancements in molecular genetics and molecular biology. Currently we know sequences of many channels, their subunit composition, stoichiometry, and transmembrane topology. These studies have demonstrated that, unlike many other classes of proteins, which are involved in fundamental physiological functions (e.g. G protein coupled receptors), ion channels do not have a common ancestor. Surprisingly, there is no correlation between functional properties and molecular organization of ion channels. For example, proton-activated channels belong to families of trimetric ASIC channels and tetrameric TRPV channels. Chloride selectivity is observed in voltage-gated ClC channels, which are homodimers with a gated pore in each subunit [2], and in pentameric glycine-gated and GABA-gated channels. On the other hand, close relatives of the latter channels, nAChR receptors, are selective for cations. Besides pentameric nAChRs, cationic selectivity is a fundamental property of tetrameric glutamate-gated channels, trimetric ATP-gated P2X channels, ASIC channels, and certain mechanosensitive channels. Both tetrameric cation-selective channels and pentameric Cys-loop channels are gated by glutamate, the major neurotransmitter in the central nervous system. The potassium selectivity is observed in five clades of potassium channels superfamily, which includes voltage-gated channels, calcium-activated channels, Kir channels, mechanosensitive two-pore (K2P) channels, and CNG/HCN channels [3]. Thus, among ion channels we can see examples of divergent evolution and examples of evolutionary homology and functional analogy. Therefore, understanding of the evolutionary history is crucially important in studies of ion channels. For example, the presence of proton-activated currents in different types of cells does not necessarily mean that these currents are mediated by the same or even evolutionary related channels. And *vice versa*, expression in a certain cell type of ion channels with close sequences does not necessarily mean that these channels have the same or even similar functional properties.

Until the pioneering publication of the first X-ray structure of a prokaryotic potassium channel, KcsA [4], 3D structures of ion channels were unknown and only indirect evidences about some features of their spatial structures were available. The lack of experimental high-resolution 3D structures greatly hindered research in the field of ion channels, including analysis of their origin and evolution. The reason for rather late appearance of 3D structures of ion channels is problems of crystallization of proteins that have both membrane-spanning and water-exposed parts. For example, MacKinnon and colleagues used bioinformatics tools to find extremely stable potassium channel in thermophilic bacteria and removed flexible cytoplasmic segments [4]. For this and subsequent seminal crystallographic studies of ion channels Roderick MacKinnon was awarded the 2003 Nobel Prize in chemistry.

Since then, many 3D structures of ion channels have been published. Most of the structures were obtained by the X-ray crystallography, but recently the amazing progress in the Cryo-EM methodology has provided high-resolution structures of different channels including open and closed states of the TRPV1 channel [5] and glycine receptor [6]. A complex of heteromeric eukaryotic calcium channel with ancillary subunits is now available [7]. The ongoing revolution in structural studies of ion channels advances research in different fields, in particular, molecular evolution of ion channels. Below we describe several representative examples.

2. Multi-domain organization of ion channels

The basic ion-conducting function of ion channels dictates that they have a transmembrane pore-forming domain. This domain is usually assembled from different subunits, which surround the central ion-conducting pore. In most of the channels the pore is lined by alpha-helical segments, but some channels have the beta-barrel structures. 3D structures show different organization of the pore-forming domains (**Figure 1**). The big variations in the number of transmembrane subunits, transmembrane topology and other structural peculiarities indicate that the domains have different evolutionary origins. In other words, various pore domains are examples of analogies rather than homologies. This conclusion is obvious in view of high-resolution structures, but before such structures become available, discrimination between analogies and homologies was by far not a trivial task.

An instructive example is the discovery of evolutionary origin of ionotropic receptors of glutamate (for review see [8]), which is the most widespread excitatory neurotransmitter in the central nervous system of vertebrates. The physiological characteristics of the glutamate receptors are similar to those of nicotinic acetylcholine receptors. Both receptor classes are ligand-gated channels, which are permeable to potassium, sodium, and, to some extent, calcium. Both receptor classes are involved in the fast synaptic transmission. In vertebrates glutamate is responsible mainly for excitation in the CNS and acetylcholine mainly provides excitation in skeletal muscles. In contrast, insects have cholinergic transmission in ganglia and glutamatergic neuromuscular transmission. These indirect evidences suggested a common origin for the superfamily of ligand-gated ion channels, including acetylcholine receptors and glutamate receptors. Comparison of the amino acid sequences of ionotropic acetylcholine and glutamate receptors reinforced this view. Indeed, in both types of receptors the neurotransmitter molecules interact with the N-terminal parts of the proteins, which are located extracellularly. Analysis of the hydrophobicity profiles revealed four putative transmembrane segments in both channel proteins. Furthermore, according to mutational data the ion selectivity and interactions with channel blockers in both channel types are controlled by residues belonging to the second potentially transmembrane segment. Thus, the idea of evolutionary homology between ionotropic receptors of glutamate and acetylcholine was supported by solid evidences [8].

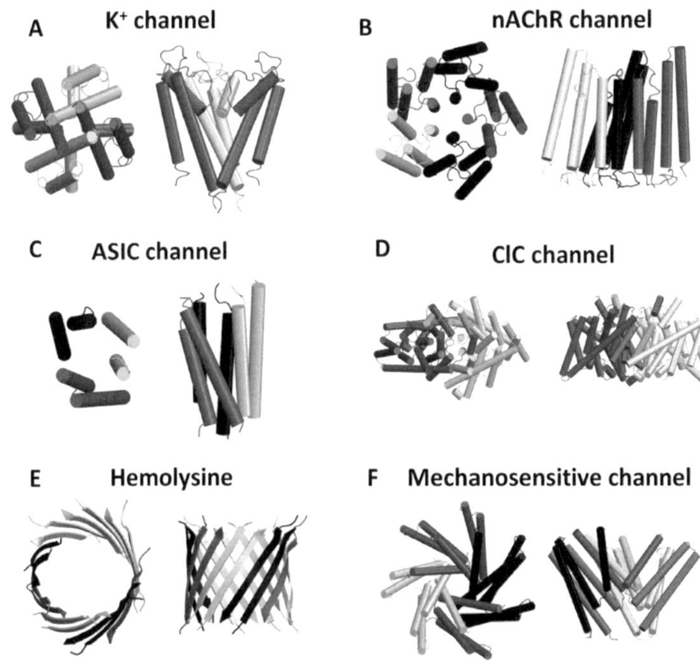

Figure 1. Diversity of pore-forming domains in ion channels. Extracellular (left) and side (right) views. (A) A potassium channel representing the family of tetrameric P-loop channels. (B) $GABA_A$ receptor channel representing pentameric channels. (C) Trimeric acid-sensing channel. (D) Dimeric chloride channel. (E) Beta-barrel structure of the hemolysin channel-forming toxin. (F) A mechanosensitive channel.

However, this idea turned out to be wrong. Increasing data on location of individual amino acid residues provided evidence that the M2 segment of the glutamate receptor does not span the entire membrane, but forms a membrane-reentrant loop both ends of which are exposed to the cytoplasm [9]. Since the transmembrane topology is one of the most conserved characteristics of membrane proteins, the hypothesis on homology between ionotropic receptors of acetylcholine and glutamate was rejected [10]. On the other hand, it is well known that the voltage-gated potassium, sodium, and potassium channels have extracellular membrane-diving loops (P-loops). Another critical series of studies demonstrated that, unlike pentameric receptors of acetylcholine, GABA and glycine, glutamate receptors have four subunits, and by this characteristic they are close to tetrameric potassium channels [11]. The final proofs were provided by the discovery an evolutionary transitional channel type in prokaryotes, which are usually difficult to find among presently existing organisms [12]. The discovery was a potassium channel activated by glutamate. This protein (named GluR0) has a ligand-binding domain, which is highly homologous to the ligand-binding domains of eukaryotic glutamate receptors. The selectivity filter of the GluR0 channel has the TVGYG sequence, which is a fingerprint of potassium channels. Importantly, the GluR0 receptor was first found by searching the database of protein sequences and then was studied experimentally [12]. Thus, glutamate receptors and voltage-gated channels inherited the pore domains from a common ancestor. This conclusion is absolutely clear from comparing 3D structures, which are available now (**Figure 2**).

However, large efforts were required to draw this conclusion before 3D structures become available.

The ligand-binding domain of glutamate receptors also has homologs among bacterial proteins (**Figure 2B**), which are glutamate-binding periplasmic proteins. Their crystal structure

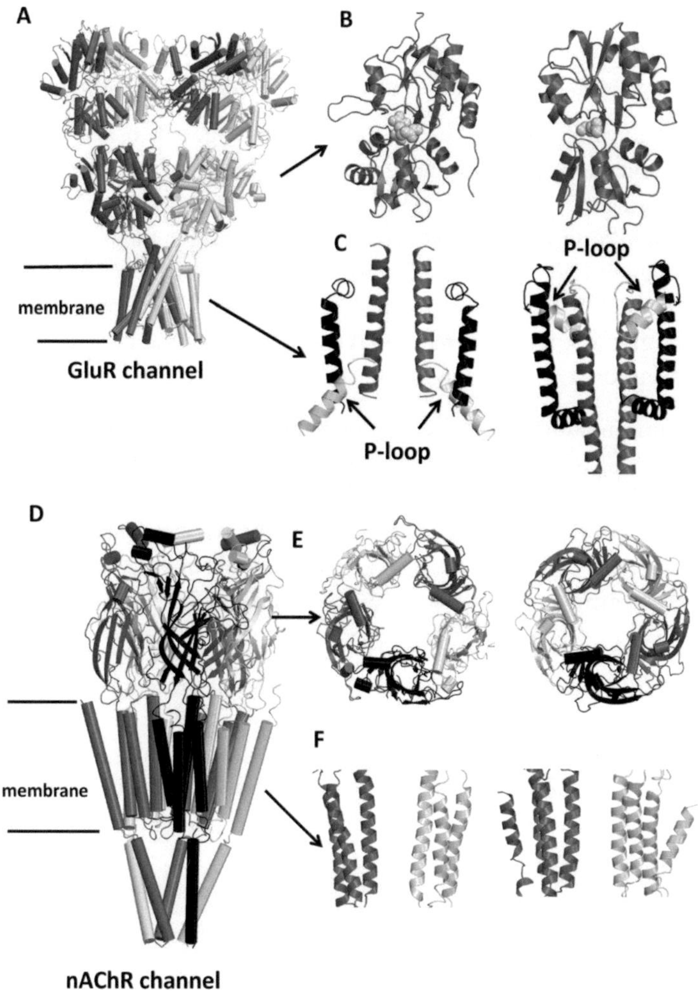

Figure 2. Different organization of glutamate (A–C) and acetylcholine (D–F) receptors. (A) In the ionotropic glutamate receptor four extracellular N-terminal domains (top), four extracellular glutamate-binding domains (middle) and a single transmembrane pore-forming domain (bottom) are connected by flexible linkers. (B) Glutamate binding domains in an ionotropic receptor (left) and in a bacterial non-channel glutamine-binding protein (right) are structurally similar. Ligands are space-filled. (C) Architecture of the pore-forming domains (only two subunits are shown for clarity) is similar in a glutamate-gated channel (left) and a voltage-gated channel (right). Note the opposite orientation of the domains: The P-loop dives into membrane from the cytoplasm in glutamate receptor and from the extracellular space in the voltage-gated channel. (D) In the nicotinic acetylcholine receptor the extracellular ligand-binding domain (top), transmembrane pore-forming domain (middle) and intracellular domain (bottom) are connected by flexible linkers. (E) Ligand-binding domains in the nicotinic acetylcholine receptor (left) and in a non-channel acetylcholine binding protein (right) are structurally similar. (F) Pore-forming domains of the cation-selective nicotinic acetylcholine receptor (right) and in the anion-selective $GABA_A$ receptor (left) are similar.

has been determined [13] and found to be remarkably similar to the glutamate binding domain in eukaryotic ionotropic receptors [14]. Probably, it is the extracellular localization of the ligand-binding domain that determines the "inverted" transmembrane topology of glutamate receptors as compared to voltage-gated channels. Nicotinic acetylcholine receptors belong to the structurally different family of so-called Cys-loop pentameric receptors, which also include channels gated by GABA, glycine and serotonin (**Figure 2**). Pore forming domains of these channels are markedly similar [15–17]. Neurotransmitter-binding domain of nicotinic acetylcholine receptors also originated from proteins that are not ion channels [18] (**Figure 2E**).

Like other ion channels of modular architecture, voltage-gated ion channels have the pore domain and four voltage-sending domains, which are believed to be of independent evolutionary origin. Indeed, some voltage-dependent enzymes (phosphatases) have voltage-sensing domains that are similar to those in ion channels. Interestingly, fusion of the voltage-sensing domains from the evolutionary unrelated phosphatase (a marine invertebrate) and a simple viral potassium channel that lacks any voltage-sensing or ligand-binding domains produced a functional voltage-gated potassium channel, which combines the pore properties of the contributing channel and the voltage dependence of the contributing phosphatase [19].

Besides the pore-forming domains and domains, which are responsible for the channel activation by specific stimuli (e.g., voltage or ligands), there are other domains involved in the channel regulation (**Figure 2A** and **D**). Regulatory functions may be performed not only by domains, but also by ancillary subunits, which are tightly associated with the channel proteins. The regulatory domains or subunits are typical characteristic of eukaryotic channels, which are involved in complex physiological systems and multiple interactions with other proteins.

Thus, various ion-channel proteins of modular organization may be assembled from domains, which originated from evolutionary and functionally distant protein families. Growing knowledge on the domain organization of ion channels shows that existent classifications of ion channels (e.g. according to the gating mechanism or the principle permeant ion) does not reflect their evolutionary history. Evolutionary studies would benefit from classification that also involves domain architecture, which is evolutionary much more conserved than physiological, biochemical or pharmacological properties.

3. Evolutionary changes in domains can govern specific channel properties

Certainly, evolutionary changes are not limited to domain organization. Homologous ionotropic receptors diverge the sensitivities to specific ligands as a result of local changes in the ligand-recognition domains. An interesting example is TRPV channels, which share with voltage-sensing P-loop channels folding of the pore domain and four "voltage-sensing" domains with apparently very similar structural organization. However, TRPV channels are not sensitive to voltage and can be activated by various stimuli, including temperature, pH, and ligands that bind to distinct sites [20]. Unexpectedly, these specific features appeared in evolution without incorporation of specific domains. Capsaicin, an active component of chili pepper, and related compounds bind in the interface between the "voltage-sensing" domain and the pore domain. Sensitivity of TRPV channels to pH is due to protonation of several residues in the extracellular loops, which have nothing in common with proton-binding domains in proton-gated ASIC channels. Thus, TRPV channels are an example, where specific properties appeared without incorporation of additional domains.

One of the most important features of pore-forming domains is their ion selectivity. Usually the selectivity is governed by a local region in the narrow part of the pore, which is named the selectivity filter. This rather small part of the channel allows very fast passing of specific types of ions and rejects other ions. Concrete mechanisms of selectivity are different, but in any case, they involve specific interactions of the permeant ions with the pore-facing amino acids.

Classical example is opposite selectivity (cationic or anionic) within the family of Cys-loop channels. The 3D structures of the pore domain are very similar (**Figure 2F**), but acetylcholine and 5-hydroxytryptamine receptors are cationic channels, whereas Glycine and $GABA_A$ receptors are anionic channels. Experiments demonstrated that just few mutations can convert the cationic selectivity to the anionic one and *vice versa*. In these channels, which discriminate anions and cations by their charge, the selectivity mechanism is realized mainly through electrostatic interactions with the rings of amino-acids in the internal and external ends of pore-lining helices.

The big and diverse family of P-loop channels includes transmembrane proteins, which permeate different types of cations. The group includes potassium, sodium, calcium channels and less selective glutamate-gated and TRP channels. Multiple studies strongly suggest that the pore-forming domains of all these channels have a common ancestor from which the folding is inherited (see above section), while specific families evolved through gene duplication and gene divergence [21]. According to 3D structures, in all these channels the ion selectivity is governed by rather small number of pore-lining residues downstream from the P-loop turns (**Figure 3**). Potassium selectivity is governed by the highly conserved VGYG motif found in various organisms including prokaryotes (**Figure 3A**). Four potassium binding sites are present in the selectivity filter (**Figure 3B**). Unlike potassium channels, in which high selectivity is achieved due to tight binding of potassium ions to the backbone carbonyls, in sodium and calcium channels the selectivity is achieved by side chains of residues, which are located in the same region. Such evolutionary divergence from a common ancestor, which was driven by the necessity of more complex electrophysiological signaling, is accompanied by local refolding of the selectivity-filter region, which is seen in 3D structures as the appearance of so-called P2 helices in sodium and calcium channels (**Figure 3C**). However, evolution of eukaryotic sodium channels from their prokaryotic ancestors includes not only replacements in the selectivity filter. Although experimental high-resolution structures of eukaryotic sodium channels are still lacking, analysis of various data suggests deletions in non-matching positions of the four repeats in the selectivity-filter region [22, 23] (**Figure 3A**).

The structure of the selectivity filter of glutamate receptors, which include NMDA-, AMPA-, and kainate-gated channels, is still unavailable, but mutational analysis suggests that selectivity is governed by the ring of Asn, Gln, or Arg residues [24]. AMPA channels, which have glutamine residues in the selectivity filter, are the most ancient among this group of channels; they are permeable for monovalent cations and calcium. NMDA channels, which have asparagine residues in the selectivity filter, are more selective for calcium ions and are blocked by Mg^{2+}. Both these properties determine physiological roles of NMDA receptors in synaptic regulation and plasticity in the glutamatergic system. On the other hand, in the GluR2 subunit

A

```
P-loop
Kv1.2          DAFWWAVVSMTT VGYGDM
NavMs          KSLYTLFQVMTL ESWSMG
Nav1.4 I       WAFLALFRLMTQ DYW-EN
Nav1.4 II      HSFLIVFRILCG E-WIET
Nav1.4 III     LGYLSLLQVATF KGW-MD
Nav1.4 IV      NSIICLFEITTS AGW-DG
Cav1.1 I       FSMLTVYQCITM EGWTDV
Cav1.1 II      QALISVFQVLTG EDWNSV
Cav1.1 III     SAMMSLFTVSTF EGWPQL
Cav1.1 IV      QAVLLLFRCATG EAWQEI
```

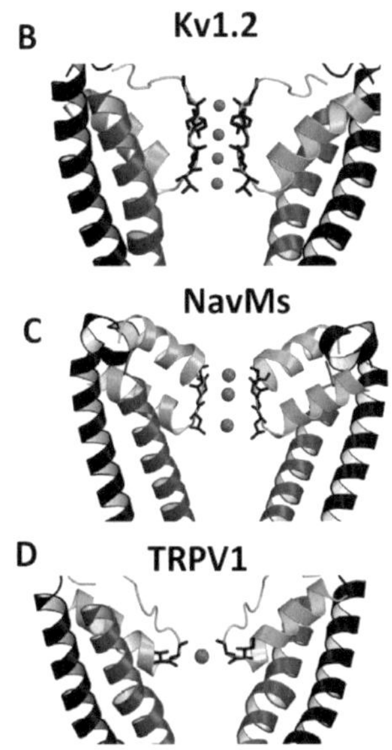

Figure 3. Selectivity in P-loop channels. (A) Sequence alignment of P-loops with the selectivity filter residues highlighted. (B–D) Available 3D structures with the ion binding sites and the selectivity filter highlighted.

of AMPA receptors Gln is replaced by Arg. This substitution completely eliminates calcium permeation. Thus, existence of three types of glutamate receptors enables different degree of coupling between electrical synaptic signaling due to permeability for sodium and potassium ions and calcium signaling, which regulates various biochemical processes within the neuron.

4. Importance of 3D structures for phylogenetic studies

Phylogenetic studies involving protein sequences are broadly used to understand molecular evolution. The first and most critical step in these studies is multiple sequence alignment of proteins. Regrettably, in the field of ion channels the standard sequence alignment tools work only within families of closely related channels, e.g., voltage-gated potassium channels [21]. However, attempts to align structurally conserved elements of the pore domains of potassium, calcium, sodium, glutamate-gated and TRP channels (all of which are P-loop channels) yielded contradictory results. Correct alignments, which later have been confirmed by comparing

http://dx.doi.org/10.5772/intechopen.73665

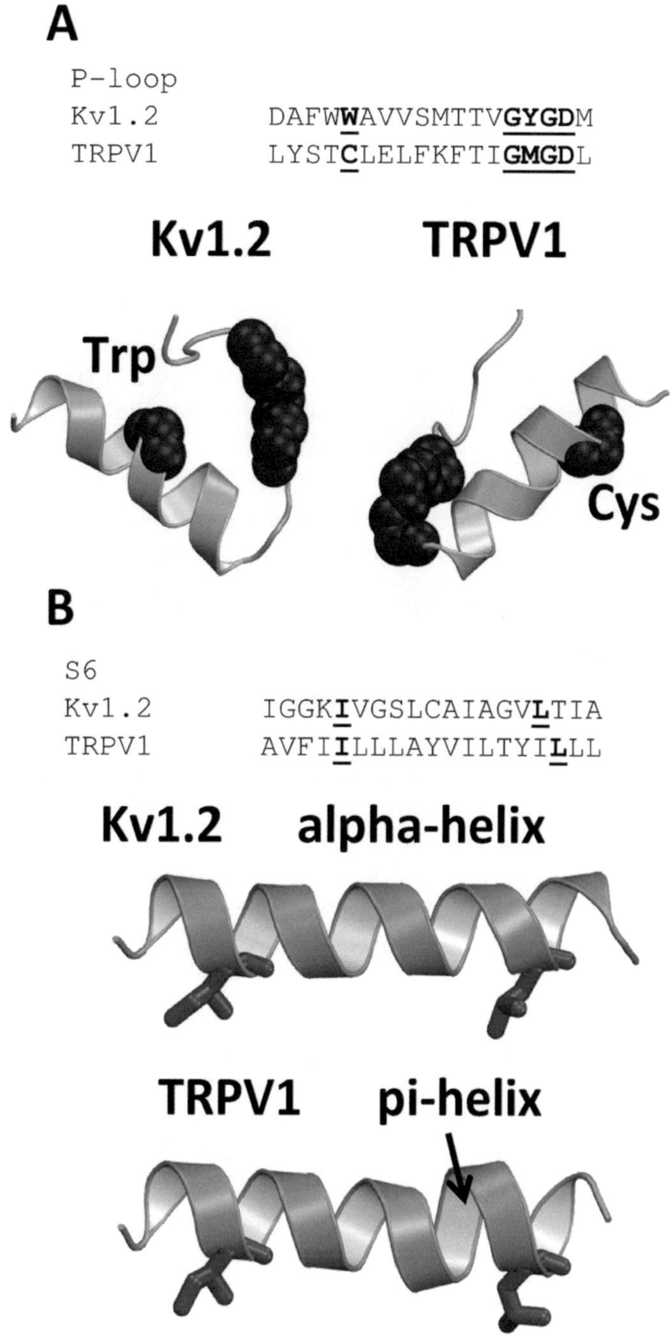

Figure 4. Mismatched sequence and 3D alignments of Kv1.2 and TRPV1 channels. (A) Alignment mismatch in P-loops. Underlined residues in matching positions have different localization in 3D structures and thus different functions. (B) Alignment mismatch in S6 segments. Underlined Leu residues in mismatching positions have the same spatial orientation due to the presence of pi-helix element.

experimental 3D structures, were elaborated by careful analysis of a large body of data on individual residues. When X-ray structures of various P-loop channels become available, the channels were found to have rather conserved folding despite vastly different sequences. Superposition of 3D structures allowed to adjust the sequence alignments, which provide much better basis for phylogenetic studies.

Alignments of residues in sequences and respective 3D structures do not necessarily coincide. For example, the sequence alignment of P-loops unambiguously shows the TVGYGDL motif of potassium channels and TVGMGDL motif of the TRPV channel in the matching positions and other positions within the P-loops do not suggest any alternative alignment (**Figure 4A**). However, in the superposed 3D structures of the TRPV and potassium channels the residues, which are in the matching positions of P-helix sequence alignment, occur at the opposite faces of the helices and thus would play different roles in the protein folding and function. In contrast, the 3D alignment, which is proposed to maximize the 3D similarity of the P-loops, shows the above residues in mismatching positions [25]. Thus, during evolution homologous residues may have changed their role and disposition, whereas the general domain folding did not change. The evolutionary mechanisms of such changes are unclear.

Transmembrane segments of ion channels are usually alpha-helices because in this secondary structure the polar groups of the backbones are hidden from the lipid environment. On the other hand, probability of possible evolutionary changes within alpha-helical structures is relatively small. Indeed, any insertion/deletion within a helical structure results in a big reorientation of other helical residues. Such reorientations would affect folding-stabilizing intersegment contacts and the exposure of functionally important groups into the aqueous pore or lipids. Therefore, the chances of acceptance of respective insertions/deletions during evolution would be small because so dramatic structural changes would result in the loss of the channel function. However, there are structural mechanisms that may provide tolerance of helical structures to insertions/deletions. The most common helical structure is an alpha-helix, which has H-bonds between residues in positions i and $i + 4$, but there are other types of helices, in which H-bonds are formed between positions i and $i + 3$ or i and $i + 5$. These secondary structures, which are called 3–10 helices and π-helices, respectively, are energetically less stable than alpha-helices and therefore are found predominantly as short segments. An insertion in an alpha-helix, which is involved in multiple intersegment contacts, would typically result in the appearance of a π-helix element (local bulging) without dramatic reorientation of residues at both sides of the insertion. An example of such a change is seen in the pore-lining S6 segment of the TRPV channel (**Figure 4B**). It was the analysis of 3D structures that allowed to reveal such changes. Analogously, deletion may results in local changes with appearance of 3–10 helices.

5. Importance of residue conservation in view of 3D structures

Multiple sequence alignments reveal conserved and variable residues. Conserved residues may be indispensable for protein folding or have crucial roles in the protein function. But it is only the 3D view that allows to really understand specific role of conserved residues. An appealing example is exceptionally conserved tryptophans in the selectivity filter region in sodium and calcium channels. It was proposed that the tryptophans are involved in the folding stabilization [26], but it is the X-ray structure of a bacterial sodium channel [27], which has revealed atomic details of H-bonds that stabilize the folding (**Figure 5B**). Tryptophan is an amino acid with the largest side chain, which can also donate an H-bond. These two factors allow tryptophan residues to simultaneously participate in a large number of specific interactions. That is why tryptophans are often found to have important structural roles as folding stabilizers. For example, in the pore helix of potassium channels side chains of two adjacent tryptophans are oriented toward neighboring segments and form multiple contacts, thus forming cyclic motifs (**Figure 5A**) in the 3D structure [25].

Asparagine residues in the middle of the pore-lining helices are highly conserved in TRPV, sodium and calcium channels. Engineered substitutions of the asparagines in sodium and calcium channels affect gating properties as well as interactions with pore-targeting ligands. Available 3D structures of sodium and calcium channels show involvement of the asparagines in inter-segment contacts, but do not allow to make an unambiguous conclusion about their functional roles. A modeling study suggests that these asparagines stabilize the open channel state [28]. Another example of highly conserved residues is positively charged arginines or lysines in every third sequential position of some transmembrane helices in voltage-gated ion channels. These positively charged residues are "signatures" of the S4 helices within the voltage-sensing domains of ion channels. At the negative resting membrane potential, the positively charged S4 helices are attracted to the cytoplasmic side of the membrane and the

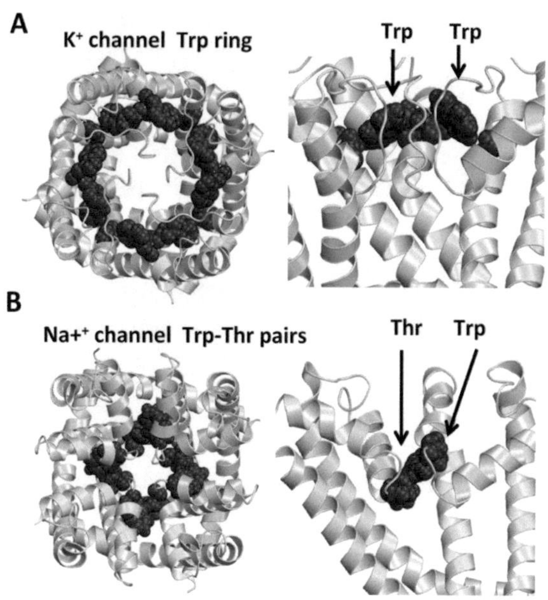

Figure 5. Conserved tryptophan residues as folding stabilizers. (A) Two Trp residues in the pore helix of potassium channel (see **Figure 3A**) form multiple contacts with neighboring segments and create the ring stabilizing the tetrameric structure. (B) Conserved Trp residues in the selectivity filter of sodium and calcium channels (see **Figure 3A**) form intersubunit contacts with Thr residues in the pore helix and provide spatial stabilization of the selectivity filter folding.

channels are closed. Upon membrane depolarization, the S4 helices shift in the extracellular direction thus initiating the process of the channel activation (opening).

The structural and/or functional importance of conserved residues is obvious. Non-conserved residues may be not critical for folding or function, but sometimes such residues play key roles in determining specific properties of the proteins. For example, variable hydrophilic residues, such as lysines, arginines, glutamines or asparagines in a matching position may indicate exposure of respective position in the aqueous environment rather than specific functional roles. However, sometimes this conclusion may be wrong. For example, asparagine, glutamine, or arginine (N/Q/R) residues determine permeability of glutamate receptor subtypes for specific cations [24]. The mechanism of selectivity was suggested in a modeling study [29], but experimental 3D structures of the N/Q/R site are still unavailable.

6. Local adaptive changes of toxin-targeted ion channels

Evolutionary changes are observable in phylogenetic studies, but driving forces for these changes remain largely unclear. Local adaptive changes of animal ion channels, which are exposed to toxins, are easy to reveal. An interesting example of the constrained convergence during evolution is resistance of snake species around the globe, which feed on newts that possess sodium channel blocker, tetrodotoxin [30]. Some garter snake populations around the globe have evolved resistance to extremely high levels of TTX [31]. Amino acid substitutions are observed in the selectivity-filter region of sodium channels and structural explanation for the acquired resistance has been proposed [26]. Interestingly, such changes are seen in the sodium channel paralogs, which are expressed in the skeletal muscle and peripheral nervous system of snakes and thus should be exposed to the ingested tetrodotoxin. In contrast, sodium channel paralogs, which are expressed solely in the central nervous system, showed no evidence of the TTX resistance, indicating that the blood-brain barrier protects from the toxin [32]. The observed genetic changes represent only a small fraction of the experimentally validated mutations known to increase the sodium channels resistance to TTX. These results suggest that evolutionary convergence at the molecular level results from the compromise between ion channel function and resistance to toxin. Thus, the natural selection may be constrained to produce similar genetic outcomes even in independent lineages.

An example of genetic adaptation of large practical importance is acquired resistance of insects to sodium channel-targeting insecticides [33]. Well known insecticides such as DDT and pyrethroids are sodium channel activators, which disturb the normal processes in the nervous system. Many insect populations worldwide mutated to elaborate residence to DDT and the earlier generation of pyrethroids. Such adaptive genetic changes are called kdr (knock-down resistance) mutations. In various inset populations many kdr mutations within sodium channel are found. Additional studies involving molecular modeling, mutational analysis and electrophysiology led to discovery of two pyrethroid binding sites in insect sodium channels [34, 35]. The receptors are located within the pore domain, in the lipid-exposed interfaces between individual channel segments whose mutual disposition changes upon the channel gating. This fast adaptive evolution of ion channels presents a big economical problem and requires development of new insecticides whose action would not be abolished by known kdr mutations. Structure-based design of new insecticides is hardly possible without knowledge of the sodium channels 3D structures.

7. Conclusions

A goal of research in the fields of molecular evolution in general and ion channels in particular is to understand how functional diversity of proteins is related to genetic changes. A traditional approach uses amino acid sequences to build phylogenetic trees and relate these with known functional characteristics of ion channels. Obviously, the amino acid sequence determines the 3D structure and functional characteristics of a protein. Nowadays, properties of small molecules are reliably predictable from their chemical structures. However, a general approach to predict 3D structures and properties of proteins just from their amino acid sequences is still lacking. Successful predictions are based on the knowledge of proteins that have similar sequences. For example, the presence of the VGYGD motif allows to predict that respective protein is a potassium channel, but such predictions are based on the fact that many channels that have this "signature sequence" have been previously studied experimentally and were found to have the potassium selectivity. It is the knowledge of 3D structures of proteins that helps to link genetic and functional data. In this work we presented examples of how the rapidly growing body of experimental data on 3D structures of ion channels influences progress in understanding their molecular evolution.

Obviously, besides molecular evolution, knowledge of 3D structures of ion channels is important in other fields. Ion channels are among the prime targets for many different drugs and toxins. This determines large demand for practical applications of knowledge on ion channels in chemistry, medicine, toxicology, and pharmacology. Structure-based analysis of action of existent drugs and toxins and design of new channel-targeting molecules requires knowledge of 3D structures of the target proteins. Available experimental 3D structures still do not represent the large diversity of ion channels and multiplicity of their functional states (open, close, inactivated, etc.). Even nowadays any new experimental 3D structure of an ion channels is an important event in the field, which is usually reported in a high-impact journal. Comparative (homology) molecular modeling is an approach to fill the gap between the numbers of desired and available 3D structures of ion channels. A key assumption of homology modeling is that the evolutionary close ion channels have similar 3D structures. Homology modeling of an ion channel involves selections of structural templates (available 3D structures of relative proteins), sequence alignment between the templates and the query protein, computer-assisted building and optimization of the model 3D structure, and analysis of possible structural deviations of the model from the templates. The choice of an incorrect template or even one-position misalignment between the template and the query protein may result in entirely incorrect model. Understanding mechanisms of molecular evolution of ion channels is necessary to avoid such errors. This is an example of how basic evolutionary studies can be translated to practical applications.

Acknowledgements

This work was supported by State budget funding by Federal Agency of Scientific Organizations Russia (Neurophysiological regulatory mechanisms and their evolution).

Abbreviations

AMPA	α-amino-3-hydroxy-5-methyl-4-isoxazolepropionic acid [receptor]
ASIC	acid-sensing ion channel
Cryo-EM	cryo-electron microscopy
ClC	chloride channel
CNG	cyclic nucleotide-gated [channels]
GABA	gamma-aminobutyric acid
HCN	hyperpolarization-activated cyclic nucleotide-modulated [channels]
Kir	inward-rectifying potassium [channels]
nAChR	nicotinic acetylcholine receptor
NMDA	N-methyl-D-aspartate [receptor]
TRPV	transient receptor potential vanilloid [channel]
KcsA	prokaryotic potassium channel from the soil bacteria *Streptomyces lividans*
TTX	tetrodotoxin

Author details

Denis B. Tikhonov* and Boris S. Zhorov

*Address all correspondence to: denistikhonov2002@yahoo.com

I.M. Sechenov Institute of Evolutionary Physiology and Biochemistry of the Russian Academy of Sciences, St. Petersburg, Russian Federation

References

[1] Rouse SL, Sansom MS. Interactions of lipids and detergents with a viral ion channel protein: Molecular dynamics simulation studies. The Journal of Physical Chemistry B. 2015;**119**:764-772

[2] Miller C. ClC chloride channels viewed through a transporter lens. Nature. 2006;**440**:484-489

[3] Brohawn SG, del Marmol J, MacKinnon R. Crystal structure of the human K2P TRAAK, a lipid- and mechano-sensitive K+ ion channel. Science. 2012;**335**:436-441

[4] Doyle DA, Morais Cabral J, Pfuetzner RA, Kuo A, Gulbis JM, Cohen SL, Chait BT, MacKinnon R. The structure of the potassium channel: Molecular basis of K^+ conduction and selectivity. Science. 1998;**280**:69-77

[5] Cao E, Liao M, Cheng Y, Julius D. TRPV1 structures in distinct conformations reveal activation mechanisms. Nature. 2013;**504**:113-118

[6] Du J, Lu W, Wu S, Cheng Y, Gouaux E. Glycine receptor mechanism elucidated by electron cryo-microscopy. Nature. 2015;**526**:224-229

[7] Wu J, Yan Z, Li Z, Yan C, Lu S, Dong M, Yan N. Structure of the voltage-gated calcium channel Cav1.1 complex. Science. 2015;**350**:aad2395

[8] Tikhonov DB, Magazanik LG. Origin and molecular evolution of ionotropic glutamate receptors. Neuroscience and Behavioral Physiology. 2009;**39**:763-773

[9] Tingley WG, Roche KW, Thompson AK, Huganir RL. Regulation of NMDA receptor phosphorylation by alternative splicing of the C-terminal domain. Nature. 1993;**364**:70-73

[10] Hollmann M, Maron C, Heinemann S. N-glycosylation site tagging suggests a three transmembrane domain topology for the glutamate receptor GluR1. Neuron. 1994;**13**: 1331-1343

[11] Rosenmund C, Stern-Bach Y, Stevens CF. The tetrameric structure of a glutamate receptor channel. Science. 1998;**280**:1596-1599

[12] Chen GQ, Cui C, Mayer ML, Gouaux E. Functional characterization of a potassium-selective prokaryotic glutamate receptor. Nature. 1999;**402**:817-821

[13] Hsiao CD, Sun YJ, Rose J, Wang BC. The crystal structure of glutamine-binding protein from *Escherichia coli*. Journal of Molecular Biology. 1996;**262**:225-242

[14] Armstrong N, Sun Y, Chen GQ, Gouaux E. Structure of a glutamate-receptor ligand-binding core in complex with kainate. Nature. 1998;**395**:913-917

[15] Spurny R, Ramerstorfer J, Price K, Brams M, Ernst M, Nury H, Verheij M, Legrand P, Bertrand D, Bertrand S, Dougherty DA, de Esch IJ, Corringer PJ, Sieghart W, Lummis SC, Ulens C. Pentameric ligand-gated ion channel ELIC is activated by GABA and modulated by benzodiazepines. Proceedings of the National Academy of Sciences of the United States of America. 2012;**109**:E3028-E3034

[16] Althoff T, Hibbs RE, Banerjee S, Gouaux E. X-ray structures of GluCl in apo states reveal a gating mechanism of Cys-loop receptors. Nature. 2014;**512**:333-337

[17] Unwin N, Fujiyoshi Y. Gating movement of acetylcholine receptor caught by plunge-freezing. Journal of Molecular Biology. 2012;**422**:617-634

[18] Brejc K, van Dijk WJ, Klaassen RV, Schuurmans M, van der Oost J, Smit AB, Sixma TK. Crystal structure of an ACh-binding protein reveals the ligand-binding domain of nicotinic receptors. Nature. 2001;**411**:269-276

[19] Arrigoni C, Schroeder I, Romani G, Van Etten JL, Thiel G, Moroni A. The voltage-sensing domain of a phosphatase gates the pore of a potassium channel. The Journal of General Physiology. 2013;**141**:389-395

[20] Grimm C, Aneiros E, de Groot M. Dissecting TRPV1: Lessons to be learned? Channels (Austin, Tex.). 2011;**5**:201-204

[21] Strong M, Chandy KG, Gutman GA. Molecular evolution of voltage-sensitive ion channel genes: On the origins of electrical excitability. Molecular Biology and Evolution. 1993;**10**: 221-242

[22] Tikhonov DB, Zhorov BS. Architecture and pore block of eukaryotic voltage-gated sodium channels in view of NavAb bacterial sodium channel structure. Molecular Pharmacology. 2012;**82**:97-104

[23] Korkosh VS, Zhorov BS, Tikhonov DB. Folding similarity of the outer pore region in prokaryotic and eukaryotic sodium channels revealed by docking of conotoxins GIIIA, PIIIA, and KIIIA in a NavAb-based model of Nav1.4. The Journal of General Physiology. 2014;**144**:231-244

[24] Traynelis SF, Wollmuth LP, McBain CJ, Menniti FS, Vance KM, Ogden KK, Hansen KB, Yuan H, Myers SJ, Dingledine R. Glutamate receptor ion channels: Structure, regulation, and function. Pharmacological Reviews. 2010;**62**:405-496

[25] Korkosh VS, Zhorov BS, Tikhonov DB. Analysis of inter-residue contacts reveals folding stabilizers in P-loops of potassium, sodium, and TRPV channels. European Biophysics Journal. 2016;**45**:321-329

[26] Tikhonov DB, Zhorov BS. Possible roles of exceptionally conserved residues around the selectivity filters of sodium and calcium channels. The Journal of Biological Chemistry. 2011;**286**:2998-3006

[27] Payandeh J, Scheuer T, Zheng N, Catterall WA. The crystal structure of a voltage-gated sodium channel. Nature. 2011;**475**:353-358

[28] Tikhonov DB, Bruhova I, Garden DP, Zhorov BS. State-dependent inter-repeat contacts of exceptionally conserved asparagines in the inner helices of sodium and calcium channels. Pflügers Archiv. 2015;**467**:253-266

[29] Tikhonov DB, Zhorov BS, Magazanik LG. Intersegment hydrogen bonds as possible structural determinants of the N/Q/R site in glutamate receptors. Biophysical Journal. 1999;**77**:1914-1926

[30] Geffeney SL, Fujimoto E, Brodie ED 3rd, Brodie ED Jr, Ruben PC. Evolutionary diversification of TTX-resistant sodium channels in a predator-prey interaction. Nature. 2005;**434**: 759-763

[31] Feldman CR, Brodie ED Jr, Brodie ED 3rd, Pfrender ME. Constraint shapes convergence in tetrodotoxin-resistant sodium channels of snakes. Proceedings of the National Academy of Sciences of the United States of America. 2012;**109**:4556-4561

[32] McGlothlin JW, Chuckalovcak JP, Janes DE, Edwards SV, Feldman CR, Brodie ED Jr, Pfrender ME, Brodie ED 3rd. Parallel evolution of tetrodotoxin resistance in three voltage-gated sodium channel genes in the garter snake Thamnophis sirtalis. Molecular Biology and Evolution. 2014;**31**:2836-2846

[33] Dong K, Du Y, Rinkevich F, Nomura Y, Xu P, Wang L, Silver K, Zhorov BS. Molecular biology of insect sodium channels and pyrethroid resistance. Insect Biochemistry and Molecular Biology. 2014;**50**:1-17

[34] Du Y, Nomura Y, Satar G, Hu Z, Nauen R, He SY, Zhorov BS, Dong K. Molecular evidence for dual pyrethroid-receptor sites on a mosquito sodium channel. Proceedings of the National Academy of Sciences of the United States of America. 2013;**110**:11785-11790

[35] O'Reilly AO, Khambay BP, Williamson MS, Field LM, Wallace BA, Davies TG. Modelling insecticide-binding sites in the voltage-gated sodium channel. The Biochemical Journal. 2006;**396**:255-263

Chapter 15

Ontogenetic and Phylogenetic Approaches for Studying the Mechanisms of Cognitive Dysfunctions

Igor A. Zhuravin, Nadezhda M. Dubrovskaya, Natalia L. Tumanova, Dmitrii S. Vasilev and Natalia N. Nalivaeva

Additional information is available at the end of the chapter

http://dx.doi.org/10.5772/intechopen.73666

Abstract

This chapter summarizes the phylogenetic and ontogenetic approaches for studying cognitive disorders such as Alzheimer's disease. It gives an extended example of evaluation of animal behavior and brain properties using an original model of prenatal hypoxia in rats by various physiological, behavioral, immunohistochemical, molecular biological, and biochemical techniques at different stages of postnatal development, which provide a better understanding of the pathological processes in the human brain during the development of neurodegeneration.

Keywords: ontogenesis, prenatal hypoxia, Alzheimer's disease, amyloid peptide, amyloid-degrading enzymes, animal models, cognitive dysfunctions, dendritic spines, synaptopodin, learning, memory, neuronal plasticity

1. Introduction

Developing the main postulates of evolutionary physiology, an eminent Russian scientist, Leon A. Orbeli, declared that for understanding the development of any functions, they should be studied from phylogenetic (using animals of different classes and species) and ontogenetic (studying organisms in development) points of view as well as under various pathological conditions and with the help of appropriate experimental models [1]. Cognitive functions represent the highest ability of the organisms to react to various stimuli from the internal and external environment by analyzing, memorizing, and storing them for immediate response or future integration and planning of the actions. They underlie all aspects of perception, thinking, reasoning, and memory. Both in humans and animals, cognitive dysfunctions are manifested as impairment of these most complex processes in the brain. The data accumulated over decades indicate that the neural mechanisms of cognitive dysfunctions can be understood as impairment of specific neural circuits, and that their functions and dysfunctions can be influenced or altered by a variety of cognitive and pharmacological factors (for review see [2]). Cognitive dysfunctions can result from various pathological changes in the brain, including significant neuronal loss observed, in particular, in Alzheimer's disease (AD).

The effectiveness of the studies of the pathogenesis of AD and search for the strategies of its prevention and treatment depend on appropriate modeling of the pathological conditions in the brain leading to AD. Traditionally, the main focus on designing animal models of AD was related to the identification of brain areas and mediator systems related to memory. The most attention has been paid to the cholinergic system that undergoes the most significant changes in AD (for review, see [3]). Several experimental approaches were proposed to model AD using injections of muscarinic receptor antagonists [4]), disruption of the medial septal area [5], or nucleus basalis of Meynert [6]. All these manipulations resulted in reduced levels of cholinacetyltransferase (ChAT) and acetylcholinesterase (AChE) in animal brain cortex leading to impairment of learning [7]. Another model employed injections of a monoclonal antibody against growth factor receptor conjugated with saporin (192 IgG-saporin), which also resulted in the loss of cholinergic neurons and cognitive disorder [8]. Intracerebroventricular injections of streptozotocin (STZ), which inhibits insulin receptor function, were also shown to lead to cholinergic loss and neurodegeneration resulting in long-term and progressive deficits in learning, memory, and cognitive behavior [9, 10]. These studies have also provided an insight into the role of diabetes in development of the sporadic form of AD [11]. However, the data regarding the changes in the content of amyloid β peptide (Aβ), which is the major component of the senile plaques and causative molecule in AD pathogenesis [12, 13], are still contradictory [9, 14]. Also, these models do not always demonstrate properly the mnemonic deficits observed at the early stages of AD.

Another common approach for modeling AD has employed, that is, injections of Aβ into specific brain areas (for review see [15]). However, these models have serious limitations and do not always reproduce the symptoms of cognitive deficit characteristics for AD [16]. Although Aβ in rodents has a slightly different amino acid sequence compared to human and monkey peptides, it can form fibrils [17] indicating that the lack of evident amyloid deposits in rodent brain is not due directly to the specific changes in its sequence but to other factors. Because of that, most animal models of AD use transgenic mice and rats, which express human AD-related proteins such as amyloid precursor protein (APP), presenilins (PS1 and PS2), and tau protein (for review see [18]).

Studies in AD transgenic mice have provided deeper insight into the processes of Aβ formation and the role of soluble Aβ oligomers in its pathogenesis [19]. Thus, it was found that in APP transgenic mice, pathological and functional changes in the brain are observed well before the formation of amyloid plaques [20]. Although in transgenic mice expressing human APP and PS formation of amyloid plaques usually accelerates with aging compared to the wild type animals in many cases, they do not demonstrate another major feature of AD which is accumulation of neurofibrillary tangles composed of tau protein. To overcome this problem, a triple transgenic mouse model (3xTgAD) was designed expressing human APP_{Sw} (containing the Swedish mutation leading to early onset AD), as well as mutated PS1 (M146 V) and tau protein (P301L) [21]. This model still is most often used in AD research. Apart from accelerated formation of amyloid plaques and neurofibrillary tangles, these mice demonstrate other pathological and behavioral feature characteristics of AD, for example, synaptic impairment and memory deterioration [22]. For elucidating the role of aging in development of sporadic AD, a model of senescence accelerated mice has been developed, which demonstrates Aβ deposits and cognitive decline as early as 6 months of age [23].

Although transgenic mouse models have prevailed over the last two decades, transgenic rat models are also becoming of more common use [24]. Despite significant physiological limitations of transgenic animals from lower phylogenetic species transgenic insects, for example, *Drosophila* [25] and worms, for example, *C. elegans* [26] have proved useful for deciphering the role of certain molecules in the development of neurodegeneration and testing various potential drugs. However, mammals still represent the main classes of animals for designing AD models since with aging some of them also develop features of neurodegeneration similar to humans [27], and recently, a transgenic mini-pig model has been approved [28].

Despite the significant data about pathogenesis of AD that have been obtained using various animal models, their major pitfall is related to the limited suitability for studying the molecular-cellular mechanisms of AD at the earliest stages of the disease when cognitive deficit is not yet accompanied by accumulation of amyloid plaques and massive cell death, but is determined by the first disruptive events in regulation of cellular interactions [29]. Also, they are mostly modeling rare genetic forms of AD, while most of the cases are sporadic late onset. Moreover, none of the therapeutic strategies based on these studies led to a successful anti-AD drug, although the amyloid cascade hypothesis underlying them has proved to be sound [30]. This can be partially explained by insufficient knowledge accumulated to date about the normal physiological role of APP and Aβ itself [31] as well as by the lack of studies modeling the early stages of AD pathology [32]. Studying mechanisms of cognitive deficit at early stages of the disease is of particular importance both from the diagnostics point of view and for design of preventive therapy. This dictates the necessity of appropriate zootropic models of experimental synaptopathies allowing to analyze the molecular and cellular bases of such conditions as mild cognitive disorder (MCI), which precedes the development of AD without modulating AD gene expression in experimental animals, and to study them at various stages of ontogenesis.

2. Cognitive functions after prenatal hypoxia

For modeling early pathologic changes of cognitive functions, we have developed a model of prenatal hypoxia in the period of the most active brain formation in rat embryogenesis. For this, Wistar female rats on the 14th day of pregnancy (E14) were placed for 3 h in a hypoxic normobaric chamber where oxygen content was gradually reduced to 7% by replacing it with helium for 10 min. The control rats were kept in the chamber for the same period of time but under normal oxygen content. The offspring of control and hypoxic rats were then subjected to various behavioral tests at different stages of ontogenesis, and their brains were taken for morphological and biochemical analysis. The rats subjected to prenatal hypoxia demonstrated general retardation and delayed formation of innate motor reactions in the postnatal period, which become less pronounced compared to controls in the process of animal development [33, 34]. However, the impairment of cognitive functions in these rats was observed in various behavioral tests throughout their life span [35].

Thus, the rats subjected to prenatal hypoxia demonstrated reduced ability to learn a new instrumental reflex such as pushing a piston inside a narrow tube. On postnatal days 20–30, the number of rats capable of learning this reflex was 50% compared to 70% in the control group. At a more advanced age (3 months old), the number of hypoxic rats capable of pushing a piston with fixed duration was 46%, while in the control group, it was about 73% [36]. Analysis of long-term memory and retrieval of the instrumental reflex after a prolonged interval (5 weeks after initial training) demonstrated that control rats were able to remember the reflex with prolonged pushing of the piston, while the rats subjected to prenatal hypoxia returned to the level of performance before the initial testing. Further training of these animals allowed them to relearn the reflex and reach the level before the interval, while the rats from the control group were able to improve their performance further. These data testify to a significant memory deficit in rats subjected to prenatal hypoxia already at the age of 3 months.

Testing the short-term memory in a radial two-level maze has revealed that adult rats subjected to prenatal hypoxia also had a significantly lower number of correct visits to the arms with unchanged average time spent in them ($p < 0.01$), which testified to the impairment of their short-term memory. In the novel object recognition test, both young and adult rats subjected to prenatal hypoxia had impaired ability to discriminate the new and old objects, both 5–10 min after the first training (short-term memory) and 60 min or 24 h after it (long-term memory) [37].

All these data allowed us to conclude that the model of prenatal hypoxia provides well-reproducible changes of cognitive functions in the postnatal ontogenesis of rats. It allows studying at the molecular levels, changes in brain structure and functions, which accompany cognitive dysfunctions, and also testing the efficacy of various pharmacological agents [37, 38].

3. Effects of prenatal hypoxia on the development of cortical brain areas

It is well known that various pathological factors in certain periods of prenatal development can lead to structural-functional changes in the brain. Existence of the periods of higher sensitivity of the brain to the pathological factors is based on the heterogeneity of ontogenetic development of the nervous system [39]. Any unfavorable factor in these critical periods of embryonic development can lead to structural-functional changes at all levels of brain organization. Such factors as ionizing radiation [40] and ultrasound [41] were shown to affect the generation and migration of the neuroblasts into the cortical plate leading to disruption in the formation of the neocortex and accompanied by prolonged disturbance in the regulation of motor activity and cognitive functions.

Using our model of prenatal hypoxia, we have also found that the underlying mechanism of the structural-functional changes observed by many authors in the postnatal development of animals subjected to hypoxia in various paradigms [42, 43] are related to the changes in generation and migration of neuroblasts caused by hypoxia in critical periods of embryonic development [44]. Using injections of 5′ethynyl-2′deoxyuridine to pregnant rats for labeling neurons generated on E14 or E18 in the fetuses, it was shown that in control rat pups, a majority of cells labeled on E14 were localized in the lower cortical layers V–VI, while the cells labeled on E18 were mainly found in the superficial cortical layers II–III. In postnatal development of rats subjected to prenatal hypoxia either on E14 or E18, we observed a certain degree of disruption in generation and migration of neuroblasts in the brain. However, hypoxia on these particular embryonic days affected different cell populations leading to specific patterns of cell labeling. Thus, hypoxia on E14, resulting in a decrease in the total number of labeled cells in the parietal cortex, led to an increase in the labeled neurons scattered in the superficial layers of the cortex of the pups. Although hypoxia on E18 also resulted in a decrease in the total number of labeled cells in the parietal cortex, the higher number of scattered labeled neurons was observed in the lower cortical layers. As a result, only rats subjected to hypoxia on E14, but not on E18, had impaired development of the whisker-placing reaction and reduced ability to learn reaching by a forepaw [44].

4. Changes in the structural-functional organization of the nervous tissue in postnatal ontogenesis after prenatal hypoxia

Using electron microscopy techniques, we have demonstrated that in early postnatal ontogenesis of rat pups subjected to prenatal hypoxia, there is a delay in formation of synaptic contacts in the neuropil, myelinization of nerve fibers, and differentiation of neurons at the ultrastructural levels both in the neocortex and basal ganglia [33, 34]. In particular, on postnatal days P10–30, we have observed a decrease in the total number of pyramidal neurons in layers II–III and V–VI of the brain cortex [45]. There were also changes in the ratio of pyramidal to nonpyramidal neurons in the first month of postnatal development of rats, which is characterized by intensive elimination of excessive cellular material in the brain and formation of intraneuronal contacts and new synapses in the cortical plate. The decrease in the number of pyramidal neurons has been observed only during the first month of postnatal development of rats, but not in adult animals, and also in the group of rats subjected to prenatal hypoxia on E14, but not on E18 [35, 44].

It is important to note the selective effect of prenatal hypoxia on different populations of cells in the brain cortex of rats. Thus, on P10–20, the rats subjected to prenatal hypoxia on E14 have a decreased number of large pyramidal neurons in layers V–VI of the neocortex due to impaired migration of neuroblasts, which forms this cell population. Application of hypoxia on E14 coincides with the period of generation of the first cells of the cortical plate, which later produce corticofugal afferents and serve as the basis for formation of cortical minicolumns. Impaired neuroblast migration in embryogenesis results in scattering of the majority of the pyramidal neurons, which should form layers V–VI of the cortex outside this area, and are eliminated on P10–30 [44]. In the period of P20–30, in animals subjected to prenatal hypoxia on E14, we have also observed a decrease in the number of small pyramidal neurons in the layers I–III and of nonpyramidal cells (interneurons) along with the total decrease of cell density in the neocortex [45].

By P60, after elimination of excessive cellular material and temporal elements, such as subplates [46], there were no significant structural differences and cell composition between control rats and rats with compromised embryonic development. However, prenatal hypoxia in the period of formation of the first elements of the cortical plate (E14) impaired formation of cortical minicolumns in postnatal ontogenesis [44], which is important for the development of proper neuronal networks and motor reactions. Hypoxia applied to a later period (E18) is not that critical for the formation of brain cortex and does not induce significant alterations in its ultrastructure. Some changes in cell composition have also been observed in the dorsal hippocampus of animals, subjected to prenatal hypoxia. The CA1 area was characterized by a lower level of neurodegenerative changes such as a decrease in the number of neurons in the pyramidal layer and an increase in the number of neurons with retracted apical dendrites. Moreover, changes in the ratio of various cell types and their delayed death have been observed only on P20 [47].

5. Role of caspases in structure-function changes in the brain after prenatal hypoxia

There is a good reason to believe that the main cause of the changes in the total density of cell distribution and of their composition after prenatal hypoxia is related to the increase in the elimination of cells caused by impaired migration of neuroblasts during embryogenesis rather than direct cell death caused by hypoxia. The data of our studies testify to upregulated expression and activity of caspase-3 and an increased number of neurons with higher expression of proapoptotic protein p53 in the cortical brain areas in rats subjected to prenatal hypoxia [45, 48], which can be interpreted as induction of cell death via caspase-dependent apoptosis. Although the molecular mechanisms of apoptosis are complex and still far from being fully understood, it is well established that activated caspases affect various proteins in the cell cytoplasm, including cellular proteases which degrade structural and regulatory proteins at the very last stages of apoptosis (for review see [49]). There are data that caspase-3 and caspase-8 are activated in the brain after hypoxia or ischemia [50, 51]. It was suggested that activation of caspases (in particular, of caspase-3) in pre- and postsynaptic terminals leads to proteolysis of various synapse-associated proteins and impairment of neuronal plasticity [49]. Systemic impairment of neuronal contacts, including axo-dendritic, observed in many pathologies and in AD, are also believed to be related to the activation of caspases due to apoptosis induced by accumulation of Aβ [52]. This concept allowed us to consider a possibility of compensating the pathology by using various caspase inhibitors.

The data obtained in our studies demonstrate that caspase-3 regulation in early postnatal ontogenesis is different in animals with normal embryonic development and subjected to prenatal hypoxia [53]. Thus, i.v. administration of inhibitors (Z-DEVD-FMK or Ac-DEVD-CHO) on 18–25 days after birth inhibited caspase-3 enzyme activity in the brain cortical structures during 3 days after the injection. However, while in animals subjected to prenatal hypoxia and characterized by increased endogenous caspase-3 activity 1–3 days after the injections, its expression and activity reduced to the control levels in the group of control rats, and administration of the inhibitors led to an increase in caspase-3 activity and expression. One month after the injection of inhibitors, the activity of caspase-3 was found to be down to the initial level characteristics of each group of animals. Administration of Ac-DEVD-CHO to mature rats (P90) with normal embryonic development also led to a decrease in the activity of caspase-3 detected 3 h after the injection. However, there was no subsequent increase of this enzyme activity on days 1 and 3 after the injection, which we observed in young intact animals. Overall, our data suggest that in young and mature rats with normal development and subjected to prenatal hypoxia, the dynamics of caspase-3 activity and properties differ significantly, especially in the period of the most intensive development of the brain [53]. Administration of inhibitors also resulted in prolonged improvement of learning and short-term memory in rats subjected to prenatal hypoxia up to the levels of control animals when tested in the two-level maze even one and half month after the injections.

6. Synaptic plasticity as the basis of adaptive potential in neuronal networks

Literature data demonstrate that dendritic spines to a great extent determine the character of cellular interactions and can be considered as a major substrate of the neuronal plasticity. The most prolonged processes involved in memory are dependent on the formation of new dendritic spines, which can rather quickly form synaptic contacts and become active during several hours [54]. The synapses themselves also undergo fast (seconds and minutes) plastic changes that affect the efficacy of synaptic transmission [55] but the formation of memory involves both fast modulations of the synapses and more slow processes of reorganization [29]. In axonal-spine synapses, the spine apparatus is involved in the process of local synthesis, posttranslational modifications, and transport of numerous synapse-associated proteins [56].

One of the marker proteins of dendritic spines is an actin-associated protein synaptopodin. The short form of this protein is localized in the spine apparatus stabilizing the cytoskeleton in the spine neck and modulating actin-based shape and motility of dendritic spines [57]. Synaptopodin is also required for cytoskeletal remodeling of the spines (changing of their size and form), and transgenic mice lacking the synaptopodin gene demonstrate short-term memory impairment, reduction of LTP, and absence of developed spine apparatus in the hippocampal dendritic spines [58]. These observations suggest that synaptopodin participates in the plasticity of neuronal networks due to its ability to reorganize the properties and distribution of labile axon-spinal neuronal contacts [56] and consolidation of memory [59].

Although disruption of synaptogenesis and formation of spine apparatus are considered among the major factors affecting neuronal plasticity, cognitive deficit, and neuronal pathologies, there are no studies evaluating the ratio of labile and stable axon-spine contacts in neuronal tissue after learning in control animals and with impaired memory. By comparing the number of labile axon-spine contacts in animals subjected to various experimental treatments, we have been able to demonstrate that both short- and long-term memory correlates with the number of synaptopodin-positive dendritic spines in the brain cortex [60]. We have shown that prenatal hypoxia in rats in the period of formation of the minicolumns in brain cortex (E14) results in a decrease in the number of synaptopodin-positive dendritic spines in the molecular layer of the neocortex and in the CA1 area of the hippocampus [61], which was accompanied by impairment of working memory. We suggest that the decrease in the number of labile synaptopodin-positive dendritic spines in the CA1 area of the hippocampus of rats subjected to prenatal hypoxia might be related to the changes in the entorhinal cortex which, in humans, is considered to be the earliest event in the development of AD [62]. According to our data, the reduction of the number of synaptopodin-positive spines along with decreased ability for learning is also observed in normally aging animals, which might be one of the reasons of cognitive dysfunctions related to advanced age, and in the sporadic form of AD [63].

7. Impairment of chemical neuronal interactions

Mechanisms of impairment of neuronal interactions caused by pathology in the embryonic period are more complex and not only involve changes in the plasticity of neuronal contacts. Literature data demonstrate that prenatal hypoxia can selectively cause disruption of various mediator systems in postnatal ontogenesis [42, 64, 65]. Using a vesicular acetylcholine transporter (VAChT) as a marker protein we have found that, in adult rats subjected to prenatal hypoxia on E14, the number of VAChT-positive cholinergic terminals, which form synapses on the bodies of the pyramidal neurons in the V–VI layers of the parietal cortex is decreased compared to control

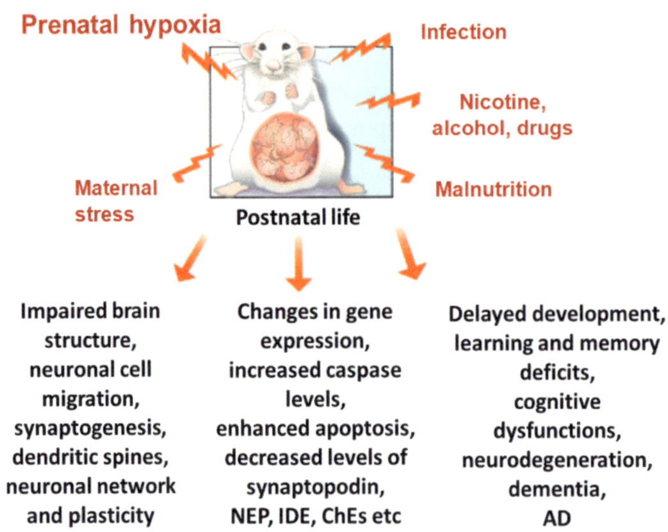

Figure 1. Effects of various types of prenatal pathology on the processes underlying development of organisms in postnatal life.

animals [66]. This also testifies to the changes in the cholinergic modulation of the cortical minicolumns in the brains of these animals, which might lead to the cognitive dysfunctions.

Comparative analysis by immunoblotting of the content of synaptophysin localized in the presynaptic terminals, independent on the nature of their mediator, as well as of the excitatory amino acid transporter (EAAT), as a marker of active glutamatergic terminals, has revealed that although in rats subjected to prenatal hypoxia synaptophysin content was not different from the controls, the EAAT levels were much higher [67]. This suggests that the intensity of glutamate release into the synaptic cleft in hypoxic animals is higher than in controls, which might provoke spontaneous epileptogenic activity and increase kindling in response to pharmacological agents and other external stimuli. These data also confirm more complex impairment in brain interconnections in postnatal ontogenesis of animals subjected to prenatal pathology. Our views on the processes underlying the changes observed in brain function after prenatal hypoxia are summarized in **Figure 1**.

8. Changes at the molecular and biochemical levels

Apart from the functional and structural changes induced by prenatal hypoxia in the nervous tissue of experimental animals, there were also significant alterations at the molecular and biochemical levels. Thus, in postnatal ontogenesis of hypoxic rats, the activity of acetyl- and butyrylcholinesterases (AChE and BChE) in the sensorimotor cortex had a significantly different dynamics compared to controls [68]. Apart from decreased activity of these enzymes during the first month of postnatal ontogenesis (and active formation of synaptic contacts), there were also significant changes in the distribution of the membrane-bound (involved in signal transduction) and soluble (participating in synaptogenesis [69]) forms of AChE. Moreover, with aging in rats subjected to prenatal hypoxia, there was an increase in the ratio of BChE in the total cholinesterase activity that could have a compensatory nature since this enzyme plays an important role in hydrolyzing various toxic agents, which might be produced by impaired brain tissue. On the other hand, increased activity of BChE in the brain is one of the characteristic features of AD and can be a marker of disruptions predisposing to neurodegeneration [70].

Analysis of the content of APP in the sensorimotor cortex also revealed different dynamics of expression of this protein in the postnatal ontogenesis of rats subjected to prenatal hypoxia [71]. While hypoxia led to an increase of the membrane bound APP at all analyzed stages of animal development, the production of its soluble forms (sAPP), which possess protective neuritogenic properties was decreased [72]. The most significant changes were observed on P10-P30 when a deficit of this neuritogenic factor might lead to disruption of formation of neuronal networks in the brain. Our data also testify that hypoxia significantly modifies the activity of α-secretase, which is important in production of the main pool of soluble APP and prevention of the formation of Aβ. The deficit of this enzyme might also lead to a decreased production of soluble AChE [73]. In general, the increase of APP content in the brain and reduction in its nonamyloidogenic processing by α-secretase after prenatal hypoxia could predispose to a shift in amyloid metabolism in the brain toward the processes initiating development of AD (**Figure 2**).

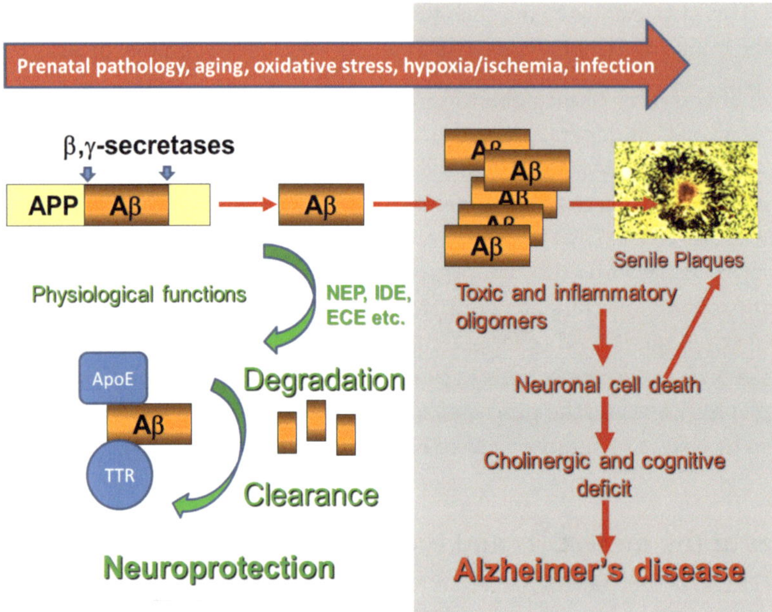

Figure 2. Schematic representation of the processes promoting the development of Alzheimer's disease. Amyloid peptide Aβ is produced from a large amyloid precursor protein (APP) after sequential cleavage by β- and γ-secretases. Aβ has a property to aggregate and form oligomers and fibrils which are toxic to the cells. In a complex with other proteins, it forms senile plaques. In the brain, there are various enzymes which can cleave Aβ, including neprilysin (NEP), insulin-degrading enzyme (IDE), and endothelin-converting enzyme (ECE). Transport proteins transthyretin (TTR) and apolipoprotein E (ApoE) are involved in Aβ clearance. Under pathological conditions, including prenatal hypoxia, and with aging, levels of NEP and IDE are reduced, and clearance of Aβ is compromised. Because hypoxia also leads to increased expression of APP, together with reduced Aβ clearance, it promotes Aβ production and accumulation resulting in the development of late onset Alzheimer's disease.

An important factor which leads to the accumulation of Aβ in the nervous tissue and causing development of the sporadic form of AD is a deficit in the activity of amyloid-degrading enzymes and impairment in the removal of this peptide from the brain (for review see [74]). According to our data, prenatal hypoxia leads to a decrease in expression of the activity of the major amyloid-degrading enzyme neprilysin (NEP) and its homolog endothelin-converting enzyme (ECE-1) [72], which could facilitate accumulation of Aβ and development of AD pathology and memory impairment. The levels of the NEP expression and activity in the brain also decrease with age in normal rats and humans as well as in AD patients [37, 72, 75]. Our experiments with administration of the NEP and ECE inhibitor phospharamidon in the cortex of intact adult rats (2×10^{-3} M, 0.25 μl per h during 28 days using a mini-pump, Alzet, USA) demonstrated a disruption of short-term memory in the radial maze and decrease in the average number of synaptopodin-positive spines in the molecular layer of brain cortex. These results were in good agreement with the data of our previous study with multiple single injections (6–8 times with one-day interval) of the NEP inhibitors phosphoramidon or thiorphan [76].

In animals subjected to prenatal hypoxia, the decrease in NEP activity was observed also in the hippocampus [37]. However, in blood plasma of such animals, we observed an increase in NEP activity compared to controls which testifies to the difference in regulation of NEP expression in the brain and peripheral tissues. Moreover, the increase of NEP expression in plasma might be a result of compensatory changes to lower the levels of brain Aβ via maintaining the existing balance between its pools in the brain and blood [77].

Analysis of expression of another amyloid-degrading protease—insulin degrading enzyme (IDE), in brain structures after prenatal hypoxia revealed its decrease in different brain structures. During postnatal ontogenesis starting from P30 and in very old rats (P600), the content of this enzyme in the cortex was lower by 40–50% and in the striatum by 30% than in the controls. The most significant decrease in IDE expression (more than by 60%) was observed in hippocampus on P20. The decrease in IDE expression was also observed in animals during normal aging and in the case of experimental diabetes [78]. Since IDE also plays an important role in insulin metabolism its deficit after prenatal hypoxia might not only lead to the pathology caused by accumulation of Aβ but also to diabetes.

Searching for the means to increase the activity of amyloid-degrading enzymes, in particular of NEP, in the brain, it was found that NEP expression can be regulated via a feedback mechanism by the C-terminal fragment of APP named AICD, which is released together with Aβ [79, 80]. It was also shown that repression of the NEP gene is epigenetically regulated by histone deacetylases (HDAC) and their inhibition by valproic acid (VA) or trichostatin A results in a significant increase in NEP mRNA and protein content as well as of enzyme activity [80]. A similar mechanism of regulation by AICD and HDAC was also shown for a transport protein, transthyretin (TTR), which is involved in removal of Aβ from the brain (**Figure 3**) [81]. Antioxidants, in particular epigallocatechin gallate (EGCG), were also shown to increase NEP expression and improve neurological deficit in Parkinson's disease and AD [82, 83]. In our experiments with animals subjected to prenatal hypoxia, it was shown that *i.p.* injections of VA or *i.c.* injections of EGCG significantly increased expression of NEP in the cortex and hippocampus bringing it up to the level of control animals [37, 63]. The increase in NEP activity correlated with an improvement of animal performance in the radial maze and with restoration of the short- and long-term memory in the novel object recognition test. Moreover, there was also an increase in the number of the labile spines in the cortex after VA injections and in the hippocampus after EGCG injections compared to animals with normal embryogenesis or injected with saline.

In the studies of the effects of hypoxia on NEP expression in human neuroblastoma cells NB7, we found that hypoxia leads to a reduction of AICD content in the cells and its binding to the NEP promoter [84], which correlated with increased expression of a number of caspases, including caspase-3, which readily cleave AICD. Addition of a caspase-3 inhibitor Z-DEVD-FMK to the cells resulted in restoration of AICD levels and NEP expression and activity which correlated with reduced amount of Aβ secreted by the cells [84]. In the rats subjected to prenatal hypoxia, we have also observed a decrease of AICD content in the brain cortex, which correlated with reduced NEP expression [48]. These studies demonstrate a link between the changes in epigenetic regulation of neuronal genes caused by prenatal hypoxia, the level of neuronal plasticity and activation of caspases, which can degrade not only transcription factors like AICD but also cytoskeletal and synaptic proteins. Further studies using our model of prenatal hypoxia will allow us and others to get a deeper insight into the mechanisms of regulation of cognitive functions at the molecular levels, which could provide a basis for design of preventive measures and therapy of cognitive disorders.

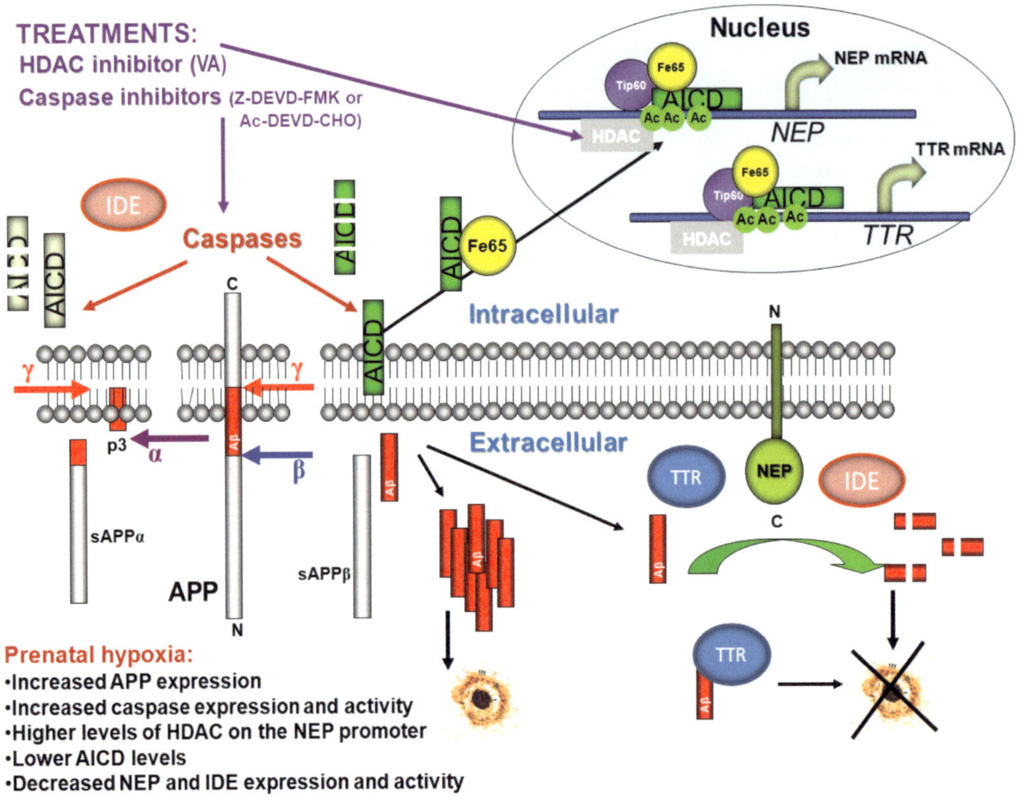

Figure 3. Effects of prenatal hypoxia on regulation of amyloid-clearing proteins and possible ways of their pharmacological up-regulation. APP is metabolized in cells via two distinct amyloidogenic and nonamyloidogenic pathways. In the amyloidogenic pathway, APP is first cleaved by β-secretase releasing a soluble ectodomain (sAPPβ) and the C-terminal fragment CTF99. The latter is cleaved by a multiprotein complex, γ-secretase, which includes presenilin-1, generating the transcriptional regulator known as the APP intracellular domain (AICD), and amyloid-β peptide (Aβ). In the nonamyloidogenic pathway, the APP molecule is first cleaved by an α-secretase within the Aβ-domain releasing a soluble ectodomain sAPPα and the C-terminal fragment CTF83. Proteolytic cleavage of CTF83 by γ-secretase also releases AICD and a short p3 fragment whose functions are still unknown. The soluble APP ectodomains, sAPPα and sAPPβ, have neuroprotective properties. AICD produced in the nonamyloidogenic pathway is cytoplasmic and readily cleaved by IDE, caspases or other proteolytic enzymes. It can also bind various proteins regulating their properties. The AICD fragment produced in the amyloidogenic pathway together with a stabilizing protein Fe65 and in a complex with other factors (including the histone acetyltransferase, Tip60) binds to an RNA polymerase mediator complex subunit Med12 acting as a transcription factor competing with histone deacetylases (HDAC) in regulation of variety of genes, including the amyloid-clearing proteins NEP and TTR allowing cells to control Aβ levels. After prenatal hypoxia in the brain, there is an increase in APP levels; however, the activation of caspases significantly reduces AICD content leading to decreased NEP expression and activity. It also results in higher levels of HDAC binding to the NEP promoter reducing its availability for AICD binding. Prenatal hypoxia also leads to reduced levels of IDE in the brain and to some changes in TTR expression as such shifting the amyloid balance toward accumulation of Aβ and neurodegeneration. Administration of an HDAC inhibitor valproic acid (VA) allows binding of AICD to the NEP promoter and restoration of NEP expression and activity. Caspase inhibitors also facilitate this process by restoring levels of AICD and protecting the brain from cognitive decline and neurodegeneration.

9. Concluding remarks

Despite the difference in the mechanisms of genesis of cognitive dysfunctions observed after prenatal pathology and development of such neurodegenerative disorders as AD, both these pathologies have common features. In early postnatal ontogenesis after prenatal hypoxia and at the early stages of mild cognitive impairment, there is dysregulation of synapse-associated proteins which is accompanied by complex disruption of neuronal interactions and by a decrease in the plasticity and adaptive potential of the cortical areas of the brain. The increase in caspase expression and activity caused by prenatal hypoxia in early postnatal ontogenesis of animals subjected to prenatal hypoxia can also lead to excessive degradation of synapse-associated proteins and degeneration of the synapses, inducing cognitive dysfunctions in postnatal ontogenesis and with aging. Changes in the activity of the enzymes participating in production and catabolism of Aβ which is the major causative agent in AD can also lead to predisposition to Aβ accumulation with aging. In rats subjected to prenatal hypoxia, these changes can be observed at significantly earlier stages of development making them a useful tool for analyzing epigenetic and molecular mechanisms of brain development during postnatal ontogenesis and leading to cognitive dysfunctions. This model can also be used for testing various pharmacological agents for their efficacy to prevent or treat various neurodegenerative disorders. Overall, these studies prove that the postulate proposed by the famous physiologist and evolutionist Leon A. Orbeli about the importance of using experimental animal models at various stages of phylogenetic and ontogenetic development as the key instruments of comparative physiology is also perfectly true and effective for studying human pathologies.

Acknowledgements

This research was carried out according to the state assignment of FASO of Russian Federation (01201351571 and Mechanisms of formation of physiological functions in phylo- and ontogenesis and influence on them of endogenous and exogenous factors) and supported in parts by RFBR grants 13-04-00388 and 16-04-00694.

Author details

Igor A. Zhuravin[1,2]*, Nadezhda M. Dubrovskaya[1,2], Natalia L. Tumanova[1],
Dmitrii S. Vasilev[1,2] and Natalia N. Nalivaeva[1]

*Address all correspondence to: zhuravin@iephb.ru

1 I.M. Sechenov Institute of Evolutionary Physiology and Biochemistry, Saint-Petersburg, Russia

2 Saint-Petersburg State Pediatric Medical University, Ministry of Health of the Russian Federation, Saint-Petersburg, Russia

References

[1] Orbeli LA. Problems of the evolutional physiology of the nervous system. (in Russian). Zhurnal Evoliutsionnoĭ Biokhimii i Fiziologii 2002;**38**:400-404

[2] Andreasen NC. Linking mind and brain in the study of mental illnesses: A project for a scientific psychopathology. Science. 1997;**275**:1586-1593

[3] Schliebs R, Arendt T. The cholinergic system in aging and neuronal degeneration. Behavioural Brain Research. 2011;**221**:555-563

[4] Flood JF, Cherkin A. Scopolamine effects on memory retention in mice: A model of dementia? Behavioral and Neural Biology. 1986;**45**:169-184

[5] Pepeu G, Casamenti F, Pedata F, Cosi C, Pepeu IM. Are the neurochemical and behavioral changes induced by lesions of the nucleus basalis in the rat a model of Alzheimer's disease? Progress in Neuro-Psychopharmacology & Biological Psychiatry. 1986;**10**:541-551

[6] Casamenti F, Prosperi C, Scali C, Giovannelli L, Pepeu G. Morphological, biochemical and behavioural changes induced by neurotoxic and inflammatory insults to the nucleus basalis. International Journal of Developmental Neuroscience. 1998;**16**:705-714

[7] McDonald MP, Overmier JB. Present Imperfect: A critical review of animal models of the mnemonic impairments in Alzheimer's disease. Neuroscience and Biobehavioral Reviews. 1998;**22**:99-120

[8] Wiley RG, Oeltmann TN, Lappi DA. Immunolesioning: Selective destruction of neurons using immunotoxin to rat NGF receptor. Brain Research. 1991;**562**:149-153

[9] Prickaerts J, Fahrig T, Blokland A. Cognitive performance and biochemical markers in septum, hippocampus and striatum of rats after an i.c.v. injection of streptozotocin: A correlation analysis. Behavioural Brain Research. 1999;**102**:73-88

[10] Salkovic-Petrisic M, Tribl F, Schmidt M, Hoyer S, Riederer P. Alzheimer-like changes in protein kinase B and glycogen synthase kinase-3 in rat frontal cortex and hippocampus after damage to the insulin signalling pathway. Journal of Neurochemistry. 2006;**96**:1005-1015

[11] Salkovic-Petrisic M, Knezovic A, Hoyer S, Riederer P. What have we learned from the streptozotocin-induced animal model of sporadic Alzheimer's disease, about the therapeutic strategies in Alzheimer's research. Journal of Neural Transmission. 2013; **120**:233-252

[12] Hardy J, Selkoe DJ. The amyloid hypothesis of Alzheimer's disease: Progress and problems on the road to therapeutics. Science. 2002;**297**:353-356

[13] Walsh DM, Klyubin I, Fadeeva JV, Rowan MJ, Selkoe DJ. Amyloid-β oligomers: Their production, toxicity and therapeutic inhibition. Biochemical Society Transactions. 2002; **30**:552-557

[14] Plaschke K, Kopitz J. In vitro streptozotocin model for modeling Alzheimer-like changes: Effect on amyloid precursor protein secretases and glycogen synthase kinase-3. Journal of Neural Transmission (Vienna). 2015;**122**:551-557

[15] Chambon C, Wegener N, Gravius A, Danysz W. Behavioural and cellular effects of exogenous amyloid-β peptides in rodents. Behavioural Brain Research. 2011;**225**:623-641

[16] Puzzo D, Gulisano W, Palmeri A, Arancio O. Rodent models for Alzheimer's disease drug discovery. Expert Opinion on Drug Discovery. 2015;**10**:703-711

[17] Fraser PE, Nguyen JT, Inouye H, Surewicz WK, Selkoe DJ, Podlisny MB, Kirschner DA. Fibril formation by primate, rodent, and Dutch-hemorrhagic analogues of Alzheimer amyloid beta-protein. Biochemistry. 1992;**31**:10716-10723

[18] Dodart JC, May P. Overview on rodent models of Alzheimer's disease. Current Protocols in Neuroscience. 2005. **33**:9.22.1-9.22.16. DOI: 10.1002/0471142301.ns0922s33

[19] Lesne S, Koh MT, Kotilinek L, Kayed R, Glabe CG, Yang A, Gallagher M, Ashe KH. A specific amyloid-beta protein assembly in the brain impairs memory. Nature. 2006; **440**:352-357

[20] Domínguez-del-Toro E, Rodríguez-Moreno A, Porras-García E, Sánchez-Campusano R, Blanchard V, Laville M, Böhme GA, Benavides J, Delgado-García JM. An in vitro and in vivo study of early deficits in associative learning in transgenic mice that over-express a mutant form of human APP associated with Alzheimer's disease. The European Journal of Neuroscience. 2004;**20**:1945-1952

[21] Oddo S, Caccamo A, Shepherd JD, Murphy MP, Golde TE, Kayed R, Metherate R, Mattson MP, Akbari Y, LaFerla FM. Triple-transgenic model of Alzheimer's disease with plaques and tangles: Intracellular Abeta and synaptic dysfunction. Neuron. 2003;**39**:409-421.2

[22] Foley AM, Ammar ZM, Lee RH, Mitchell CS. Systematic review of the relationship between amyloid-β levels and measures of transgenic mouse cognitive deficit in Alzheimer's disease. Journal of Alzheimer's Disease. 2015;**44**:787-795

[23] Del Valle J, Duran-Vilaregut J, Manich G, Casadesús G, Smith MA, Camins A, Pallàs M, Pelegrí C, Vilaplana J. Early amyloid accumulation in the hippocampus of SAMP8 mice. Journal of Alzheimer's Disease. 2010;**19**:1303-1315

[24] Do Carmo S, Cuello AC. Modeling Alzheimer's disease in transgenic rats. Molecular Neurodegeneration. 2013;**8**:37. DOI: 10.1186/1750-1326-8-37

[25] Newman T, Sinadinos C, Johnston A, Sealey M, Mudher A. Using Drosophila models of neurodegenerative diseases for drug discovery. Expert Opinion on Drug Discovery. 2011;**6**:129-140

[26] Calahorro F, Ruiz-Rubio M. *Caenorhabditis elegans* as an experimental tool for the study of complex neurological diseases: Parkinson's disease, Alzheimer's disease and autism spectrum disorder. Invertebrate Neuroscience. 2011;**11**:73-83

[27] Youssef SA, Capucchio MT, Rofina JE, Chambers JK, Uchida K, Nakayama H, Head E. Pathology of the aging brain in domestic and laboratory animals, and animal models of human neurodegenerative diseases. Veterinary Pathology. 2016;**53**:327-348

[28] Holm IE, Alstrup AK, Luo Y. Genetically modified pig models for neurodegenerative disorders. The Journal of Pathology. 2016;**238**:267-287

[29] Arendt T. Synaptic plasticity and cell cycle activation in neurons are alternative effector pathways: The 'Dr. Jekyll and Mr. Hyde concept' of Alzheimer's disease or the yin and yang of neuroplasticity. Progress in Neurobiology. 2003;**71**:83-248

[30] Hardy J. The amyloid hypothesis for Alzheimer's disease: A critical reappraisal. Journal of Neurochemistry. 2009;**110**:1129-1134

[31] Nalivaeva NN, Turner AJ. The amyloid precursor protein: A biochemical enigma in brain development, function and disease. FEBS Letters. 2013;**587**:2046-2054

[32] Mufson EJ, Ikonomovic MD, Counts SE, Perez SE, Malek-Ahmadi M, Scheff SW, Ginsberg SD. Molecular and cellular pathophysiology of preclinical Alzheimer's disease. Behavioural Brain Research. 2016;**311**:54-69

[33] Zhuravin IA, Tumanova NL, Ozirskaya EV, Vasilyev DS, Dubrovskaya NM. Formation of striatum structural and ultrastructural organization in the early postnatal ontogenesis of rats subjected to altered conditions of their embryonic development (In Russian). Morfologiia. 2005;**127**:31-36

[34] Zhuravin IA, Tumanova NL, Ozirskaya EV, Vasil'ev DS, Dubrovskaya NM. Formation of the structural and ultrastructural organization of the striatum in early postnatal ontogenesis of rats in altered conditions of embryonic development. Neuroscience and Behavioral Physiology. 2006;**36**(5):473-478. DOI: 10.1007/s11055-006-0042-8

[35] Dubrovskaya NM, Zhuravin IA. Ontogenetic characteristic of behavior in rats subjected to prenatal hypoxia on the 14-th or 18-th days of embryogenesis. Neuroscience and Behavioral Physiology. 2010;**40**(2):231-238. DOI: 10.1007/s11055-009-9235-2

[36] Zhuravin IA, Dubrovskaya NM, Plesneva SA. Striatal level of regulation of learned fore-paw movements in rats. Physiological Research. 2002;**51**(Suppl 1):67-76

[37] Nalivaeva NN, Belyaev ND, Lewis DI, Pickles AR, Makova NZ, Bagrova DI, Dubrovskaya NM, Plesneva SA, Zhuravin IA, Turner AJ. Effect of sodium valproate administration on brain neprilysin expression and memory in rats. Journal of Molecular Neuroscience. 2012;**46**:569-577

[38] Zhuravin IA, Dubrovskaya NM, Tumanova NL. Postnatal physiological development of rats after acute prenatal hypoxia. Neuroscience and Behavioral Physiology. 2004; **34**:809-816

[39] Rice D, Baron S. Critical periods of vulnerability for the developing nervous system: Evidence from humans and animal models. Environmental Health Perspectives 2000; **108**: 511-533.

[40] Algan O, Rakic P. Radiation-induced, lamina-specific deletion of neurons in the primate visual cortex. The Journal of Comparative Neurology. 1997;**381**:1096-9861

[41] Ang Jr ES, Gluncic V, Duque A, Schafer ME, Rakic P. Prenatal exposure to ultrasound waves impacts neuronal migration in mice. Proceedings of the National Academy of Sciences of the United States of America. 2006;**103**:12903-12910

[42] Nyakas C, Buwalda B, Luiten PGM. Hypoxia and brain development. Progress in Neurobiology. 1996;**49**:1-51

[43] Otellin VA, Khozhaĭ LI, Gilerovich EG, Korzhevskiĭ DE, Kostkin VB, Belostotskaia GB. Damaging impacts at the critical time periods of prenatal ontogenesis as a factor modifying the cerebral structural development and postnatal behavior reactions (In Russian). Vestnik Rossiĭskoĭ Akademii Meditsinskikh Nauk 2002;**12**:32-35

[44] Vasilev DS, Dubrovskaya NM, Tumanova NL, Zhuravin IA. Prenatal hypoxia in different periods of embryogenesis differentially affects cell migration, neuronal plasticity, and rat behavior in postnatal ontogenesis. Frontiers in Neuroscience. 2016;**10**:126. DOI: 10.3389/fnins.2016.00126

[45] Vasiliev DS, Tumanova NL, Zhuravin IA. Change of the neocortex nervous tissue in rat ontogenesis after hypoxia at various terms of embryogenesis. Journal of Evolutionary Biochemistry and Physiology. 2008;**44**(3):304-315. DOI: 10.1134/S002209300803006X

[46] Del Rio JA, Martinez A, Auladell C, Soriano E. Developmental history of the subplate and developing white matter in the murine neocortex. Neuronal organization and relationship with the main afferent systems at embryonic and perinatal stages. Cerebral Cortex. 2000;**10**:784-801

[47] Zhuravin IA, Tumanova NL, Vasiliev DS. Structural changes of hippocampus nervous tissue in rat ontogenesis after prenatal hypoxia. Journal of Evolutionary Biochemistry and Physiology. 2009;**45**:156-158. DOI: 10.1134/S0022093009010165

[48] Kozlova DI, Vasylev DS, Dubrovskaya NM, Nalivaeva NN, Tumanova NL, Zhuravin IA. Role of caspase-3 in regulation of the amyloid-degrading neuropeptidase neprilysin level in the rat cortex after hypoxia. Journal of Evolutionary Biochemistry and Physiology. 2015;**51**:480-484. DOI: 10.1134/S0022093015060046

[49] Snigdha S, Smith ED, Prieto GA, Cotman CW. Caspase-3 activation as a bifurcation point between plasticity and cell death. Neuroscience Bulletin. 2012;**28**:14-24

[50] Delivoria-Papadopoulos M, Ashraf QM, Ara J, Mishra OP. Nuclear mechanisms of hypoxic cerebral injury in the newborn: The role of caspases. Seminars in Perinatology. 2008;**32**:334-343

[51] Broughton BR, Reutens DC, Sobey CG. Apoptotic mechanisms after cerebral ischemia. Stroke. 2009;**40**:331-339

[52] D'Amelio M, Cavallucci V, Middei S, Marchetti C, Pacioni S, Ferri A, Diamantini A, De Zio D, Carrara P, Battistini L, Moreno S, Bacci A, Ammassari-Teule M, Marie H, Cecconi F. Caspase-3 triggers early synaptic dysfunction in a mouse model of Alzheimer's disease. Nature Neuroscience. 2011;**14**:69-76

[53] Vasilev DS, Dubrovskaya NM, Nalivaeva NN, Zhuravin IA. Regulation of caspase-3 content and activity in rat cortex after prenatal hypoxia. Neurochemical Journal. 2016; **10**(2):144-150

[54] Zito K, Scheuss V, Knott G, Hill T, Svoboda K. Rapid functional maturation of nascent dendritic spines. Neuron. 2009;**61**:247-258

[55] Martin KC, Barad M, Kandel ER. Local protein synthesis and its role in synapse-specific plasticity. Current Opinion in Neurobiology. 2000;**10**:587-592

[56] Segal M. Dendritic spines, synaptic plasticity and neuronal survival: Activity shapes dendritic spines to enhance neuronal viability. The European Journal of Neuroscience. 2010;**31**:2178-2184

[57] Mundel P, Heid HW, Mundel TM, Krüger M, Reiser J, Kriz W. Synaptopodin: An actin-associated protein in telencephalic dendrites and renal podocytes. The Journal of Cell Biology. 1997;**139**:193-204

[58] Deller T, Korte M, Chabanis S, Drakew A, Schwegler H, Stefani GG, Zuniga A, Schwarz K, Bonhoeffer T, Zeller R, Frotscher M, Mundel P. Synaptopodin-deficient mice lack a spine apparatus and show deficits in synaptic plasticity. Proceedings of the National Academy of Sciences of the United States of America. 2003;**100**:10494-10499

[59] Asanuma K, Kim K, Oh J, Giardino L, Chabanis S, Faul C, Reiser J, Mundel P. Synaptopodin regulates the actin-bundling activity of α-actinin in an isoform-specific manner. The Journal of Clinical Investigation. 2005;**115**:1188-1198

[60] Dubrovskaya NM, Nalivaeva NN, Vasilev DS, Bagrova DI, Zhuravin IA. Mechanisms of short-term working memory deficit. In: Short-Term Memory: New Research. New York: Nova Science Publishers Inc.; 2012. pp. 155-173

[61] Zhuravin IA, Tumanova NL, Vasilyev DS. Changes in the mechanisms of brain adaption during postnatal ontogeny of rats exposed to prenatal hypoxia. Doklady Biological Sciences. 2009;**425**:91-93. DOI: 10.1134/S001249660902001X

[62] Gómez-Isla T, Price JL, McKeel Jr DW, Morris JC, Growdon JH, Hyman BT. Profound loss of layer II entorhinal cortex neurons occurs in very mild Alzheimer's disease. The Journal of Neuroscience 1996;**16**:4491-4500

[63] Zhuravin IA, Dubrovskaya NM, Vasilev DS, Tumanova NL, Nalivaeva NN. Epigenetic and pharmacological regulation of the amyloid-degrading enzyme neprilysin results in modulation of cognitive functions in mammals. Doklady Biological Sciences. 2011; **438**:145-148

[64] Louzoun-Kaplan V, Zuckerman M, Perez-Polo JR, Golan HM. Prenatal hypoxia down regulates the GABA pathway in newborn mice cerebral cortex; partial protection by $MgSO_4$. International Journal of Developmental Neuroscience. 2008;**26**:77-85

[65] Gerstein M, Huleihel M, Mane R, Stilman M, Kashtuzki I, Hallak M, Golan H. Remodeling of hippocampal GABAergic system in adult offspring after maternal hypoxia and magnesium sulfate load: Immunohistochemical study. Experimental Neurology. 2005; **196**:18-29

[66] Vasil'ev DS, Dubrovskaya NM, Tumanova NL, Zhuravin IA. Effects of transient stress on the plastisity of the cortical areas of the brain and cognitive functions in adult rats

with normal and disturbed embryogenesis. Neuroscience and Behavioral Physiology. 2015;**45**(6):643-647. DOI: 10.1007/s11055-015-0123-7

[67] Vasil'ev DS, Tumanova NL, Lavrent'eva VV, Starshinova LA, Zhabko EP, Lukomskaya NI, Zhuravin IA, Magazanik LG. The ability of NMDA-type glutamate receptor blockers to prevent the development a pentylenetetrazole kindling and morphological changes to pyramidal neurons in the mouse hippocampus. Neuroscience and Behavioral Physiology. 2015;**45**(5):528-535. DOI: 10.1007/s11055-015-0106-8

[68] Kochkina EG, Plesneva SA, Zhuravin IA, Nalivaeva NN, Turner AJ. The effect of hypoxia on cholinesterase activity in rat sensorimotor cortex. Journal of Evolutionary Biochemistry and Physiology. 2015;**51**:107-116. DOI: 10.1134/S0022093015020039

[69] Halliday AC, Greenfield SA. From protein to peptides: A spectrum of non-hydrolytic functions of acetylcholinesterase. Protein and Peptide Letters. 2012;**19**:165-172

[70] Greig NH, Lahiri DK, Sambamurti K. Butyrylcholinesterase: An important new target in Alzheimer's disease therapy. International Psychogeriatrics. 2002;**14**(Suppl 1):77-91

[71] Nalivaeva NN, Fisk L, Canet Aviles RM, Plesneva SA, Zhuravin IA, Turner AJ. Effects of prenatal hypoxia on expression of amyloid precursor protein and metallopeptidases in the rat brain. Letters in Peptide Science. 2003;**10**:455-462

[72] Nalivaeva NN, Fisk L, Kochkina EG, Plesneva SA, Zhuravin IA, Babusikova E, Dobrota D, Turner AJ. Effect of hypoxia/ischemia and hypoxic preconditioning/reperfusion on expression of some amyloid-degrading enzymes. Annals of the New York Academy of Sciences. 2003;**1035**:21-33

[73] Nalivaeva NN, Turner AJ. Does acetylcholinesterase secretion involve an ADAMs-like metallosecretase? Letters in Peptide Science. 1999;**6**:343-348

[74] Nalivaeva NN, Beckett C, Belyaev ND, Turner AJ. Are amyloid-degrading enzymes viable therapeutic targets in Alzheimer's disease? Journal of Neurochemistry. 2012;**120**(Suppl 1): 167-185

[75] Wang S, Wang R, Chen L, Bennett DA, Dickson DW, Wang DS. Expression and functional profiling of neprilysin, insulin-degrading enzyme, and endothelin-converting enzyme in prospectively studied elderly and Alzheimer's brain. Journal of Neurochemistry. 2010;**115**:47-57

[76] Dubrovskaya NM, Nalivaeva NN, Plesneva SA, Feponova AA, Turner AJ, Zhuravin IA. Changes in the activity of amyloid-degrading metallopeptidases leads to disruption of memory in rats. (In Russian). Zhurnal Vysshei Nervnoi Deiatelnosti Imeni I. P. Pavlova. 2009;**59**:630-638

[77] Guan H, Liu Y, Daily A, Police S, Kim MH, Oddo S, LaFerla FM, Pauly JR, Murphy MP, Hersh LB. Peripherally expressed neprilysin reduces brain amyloid burden: A novel approach for treating Alzheimer's disease. Journal of Neuroscience Research. 2009;**87**:1462-1473

[78] Kochkina EG, Plesneva SA, Vasilev DS, Zhuravin IA, Turner AJ, Nalivaeva NN. Effects of ageing and experimental diabetes on insulin-degrading enzyme expression in male rat tissues. Biogerontology. 2015;**16**:473-484. DOI: 10.1007/s10522-015-9569-9

[79] Pardossi-Piquard R, Petit A, Kawarai T. Sunyach C, Alves da Costa C, Vincent B, Ring S, D'Adamio L, Shen J, Müller U, St George Hyslop P, Checler F. Presenilin-dependent transcriptional control of the Aβ-degrading enzyme neprilysin by intracellular domains of βAPP and APLP. Neuron 2005;**46**:541-554

[80] Belayev ND, Nalivaeva NN, Makova NZ, Turner AJ. Neprilysin gene expression requires binding of the amyloid precursor protein intracellular domain to its promoter: Implications for Alzheimer disease. EMBO Reports. 2009;**10**:94-100

[81] Kerridge C, Belyaev ND, Nalivaeva NN, Turner AJ. The Aβ-clearance protein transthyretin, like neprilysin, is epigenetically regulated by the amyloid precursor protein intracellular domain. Journal of Neurochemistry. 2014;**130**:419-431

[82] Melzig MF, Janka M. Enhancement of neutral endopeptidase activity in SK-N-SH cells by green tea extract. Phytomedicine. 2003;**10**:494-498

[83] Weinreb O, Mandel S, Amit T, Youdim MB. Neurological mechanisms of green tea polyphenols in Alzheimer's and Parkinson's diseases. The Journal of Nutritional Biochemistry. 2004;**15**:506-516

[84] Kerridge C, Kozlova DI, Nalivaeva NN, Turner AJ. Hypoxia affects neprilysin expression through caspase activation and an APP intracellular domain-dependent mechanism. Frontiers in Neuroscience. 2015;**9**:426. DOI: 10.3389/fnins.2015.00426

Chapter 16

Appendix: The Main Problems and Methods of Evolutionary Physiology—A Lecture by Leon A. Orbeli

Additional information is available at the end of the chapter

http://dx.doi.org/10.5772/intechopen.73670

The following text is a lecture given at the First Conference on Evolutionary Physiology, March 7, 1956. It was originally published in Russian, in the book Evolutsiya funktsii nervnoi sistemy (Evolution of Functions of the Nervous System), Leningrad, 1958, pp. 7–17. This text was written in the Soviet epoch and therefore it contains sometimes necessary "reverences" in the direction of communistic ideology.

The text was translated by Andrey Polyanovsky.

The question of evolutionary physiology as a discipline in its own right was raised only in our country. This may have a whole number of explanations. First of all, some outstanding scientists in our country followed the evolutionary line in studying one or another physiological issue. The necessity of the evolutionary approach in physiology was accentuated by I.P. Pavlov, I.M. Sechenov and N.E. Vvedensky. From them, we got not only statements, but also fundamental papers, which predetermined further development of the evolutionary trend in physiology. This especially concerns I.P. Pavlov's studies of the higher nervous activity. I.P. Pavlov had good reasons to emphasize that studying conditioned reflexes is essentially studying reflexes in the process of their formation and development, in their beginnings and that, when studying a conditioned reflex activity, a researcher views the entire developmental continuum of reflex activity and, consequently, gets a chance to deduce from the history of conditioning the overall process of reflex activity formation during evolution. This has served a major impetus for physiologists in our country, rather than in any other, to choose the evolutionary road and to begin following the evolutionary principle.

Alongside the directions left by the coryphaei of experimental science, we have yet another, no less and may be even more important, explanation—the fact that all Soviet science is advancing under the guidance of the solely correct philosophical doctrine of dialectic materialism. The Marxist-Leninist doctrine obliges every researcher in the field of natural sciences to adhere to the principle that no one phenomenon can be understood without analyzing the history of its emergence, its development. It is the only historical method that enables correct understanding of the subject matter. Based on these two tenets—the requirements of the Marxist-Leninist philosophy and the directions by our great predecessors—we are approaching now the problems of evolutionary physiology.

It would be incorrect to say that wishes to study functions in their evolutionary history were not stated by other eminent physiologists. I remember that in 1908, the now late English physiologist Keith Lucas published his brochure entitled "Evolution of functions," in which he expressed bitter regret that the evolutionary principle, which proved to be extremely fruitful and widely used in morphological sciences,* was almost of no use in physiology. K. Lucas stated some considerations on the necessity of following the evolutionary principle in physiological studies and, furthermore, pointed out the difficulties physiologists encounter if they attempt to approach a study of the function from the viewpoint of its emergence and origin. In any case, our Soviet physiology can be presently characterized exactly as physiology imbued with the evolutionary principle. Nevertheless, the question does arise on whether it is enough for individual researchers working in the field of physiology simply to adhere to evolutionary theory, to reckon with evolutionary theory, to try applying the historic approach to solving one or another problem or to seek a new discipline entitled "evolutionary physiology" to appear and develop as an independent discipline within the system of physiological studies and biological sciences.

I can afford asserting that the time has come not simply for the evolutionary principle to become a guiding force in furthering physiological studies but for the independent discipline, evolutionary physiology, to arise as a result of the progress achieved by physiological science in general.

I must say that morphologists already pointed out the necessity of establishing evolutionary physiology along with evolutionary morphology. These statements belong to late A.N. Severtsov, who emphasized that it is time for physiologists to address the evolutionary matters and to found evolutionary physiology along with evolutionary morphology. The term "evolutionary physiology," still unknown in other countries, is our term, our proposition to single out evolutionary physiology as a discipline in its own right along with evolutionary morphology, evolutionary histology and evolutionary biochemistry.

I have to remind that in this case it is definitely not about tearing evolutionary physiology as an independent discipline from the rest of physiology. Not at all! We should consider it as a new, present-day stage in the development of physiology because evolutionary physiology can in no way be constructed in isolation from the rest of physiology. It should make maximum use of all that abundant material, which was obtained both by classical (medical) physiology and by general physiology, zoophysiology, comparative physiology and embryophysiology—all those branches of physiology that have been elaborated until now and have achieved enormous success. Evolutionary physiology should use all this material. It is a product of, a superstructure over, those physiological studies that are in progress now. But this does not deprive it of the right to put forward its own questions, to pose them as a leading line, and to select the material that helps solve the evolutionary problem.

What should be the tasks of evolutionary physiology? It seems to me that the two pivotal lines should underlie evolutionary physiology. K. Lucas grieved that there were no studies of the evolution of functions, and of course, the evolution of functions must be the first pivot of evolutionary physiology. We should strive to consider any function subjected to experimental investigation in terms of the history of its formation: how one or another function is formed in the process of evolution, how the intertwining of separate functions led to certain changes in

http://dx.doi.org/10.5772/intechopen.73670

each of them, resulting in the representation of each given function in certain animals exactly this way and not otherwise. This is the question of studying the evolution of functions.

Another question is whether evolutionary physiology should and could be confined to studying the evolution of functions. I do not think so. The second, no less and may be even more important task of evolutionary physiology, is supposed to be functional evolution, that is, a verification of evolutionary physiology on the basis not of morphological material and morphological methods but physiological studies.

A study of the evolution of functions will provide a certain material and open up a new avenue toward understanding why the evolutionary process was running this way, but not otherwise, and what the mainstream of the evolutionary process was in terms of the evolution of functions.

It may appear that these are one and the same. But essentially it is not true, indeed. Functional evolution is a higher stage of evolutionary physiology as compared to studying the evolution of functions. In one case, we simply track the historical developmental path of some functional relationships, while in another case we approach the understanding of what the evolutionary process is and how it formed, why the evolutionary process was running exactly this way being based on those functional rearrangements that were arising in living organisms.

Proceeding now to more specific tasks that underlie both main pivots of evolutionary physiology, we must pose a question: is it enough to study the developmental path of various functional relationships in different representatives of the animal kingdom we are dealing with, taking into account the history of the formation of these functions, or should we also set ourselves the task of elucidating the mechanisms of the evolution of functions, those specific conditions and motives that directed the process of development along one or another way? In other words, should we also study the significance of the individual factors that determined the course of evolution?

In this respect, we certainly have to reckon with the basic tenet of our science and Marxist-Leninist philosophy that organism and its environment make up a single, indivisible and interrelated whole. With this in mind and considering that the entire process of the development of different functional relationships ran in a certain environmental context, everchanging, ever affecting living organisms, it becomes clear that no one function could form and undergo any changes otherwise than under the influence of and depending on those environmental impacts that it was permanently exposed to. So, the task of studying the evolution of functions involves not only the elucidation of the course of development, developmental pattern, succession of events, but also establishing the interdependence of these events as well as the causative dependence of all changes and transformations on the environmental factors that affect living beings. We have to reckon both with the internal factors coming from the organism itself, in the form of an interaction of its separate parts, and the external factors.

As regards the problem of functional evolution, this is a much more complicated matter. Here we need to foresee the issue of what further development is supposed to be, how we could imagine the further transformation of individual functions and whole organisms under the effect of those environmental changes that are occurring right now, before our eyes, and that may become determinative for further development of functional relationships. These issues are not only of theoretical significance. Of course, their theoretical significance is quite clear to everyone and requires no proof, but I have to remind that the conditions, under which the organisms are living at present, change substantially every day. Enormous success of science and technology, that we are witnessing now and that resulted from great progress humankind has achieved in its evolution, creates by itself new living conditions, sometimes so different from the normal that they may prove to be determinative for further development of life on Earth. From this viewpoint, we have to admit that evolutionary physiology, in the sense I have stated above, is not only a theoretical but also a strictly practical science in the sense of considering all those conditions, under which we are living now and will be living in the near future, in the light of those influences that these conditions may have on the existing organisms and their offspring, be it of human beings or the animal world.

The question arises: what methods should evolutionary physiology use to embrace the tasks I have just talked about?

Over many years, I and my fellow workers have been adhering to the viewpoint that correct understanding of the evolution of functions and mechanisms underlying evolutionary changes in functions is attainable provided that one and the same researcher uses simultaneously four methods, basically different but leading to the same goal.

The first way, as is clear to everyone, is certainly **comparative physiology**—the use of comparative physiological material in order to understand how different phyletic lines evolved under different living conditions and how the same functions developed becoming more sophisticated or, quite the contrary, dropping out in some phyletic lines depending on the specific living conditions. The question then arises as to how one and the same function undergoes changes under different conditions and, on the other hand, how the initially different functional relationships converge leading to one and the same ultimate result under the influence of environmental factors.

Evolutionary physiology should definitely be based on the ready material of comparative physiology and *zoophysiology*, but on the other hand, it should develop on its own, artificially selecting those animal samples and conditions that are of special interest from the viewpoint of the evolution of functions. Here, perhaps, we have to run counter to the views of some evolutionists. We believe that it is necessary to distinguish between three disciplines. The first is zoophysiology, that is, descriptive physiology of individual animal species, an extremely important science, which is of great theoretical and even greater practical significance. We need to know all forms of life represented now on Earth. Another thing is *comparative physiology*, which chooses from the huge material of zoophysiology only certain objects and issues that allow a comparison and elucidation of certain regularities. Yet more special and, at the same time, complex demands are made by *evolutionary physiology*, which not simply compares what occurs under different living conditions but uses the experimental method to understand how these functions formed.

The second way is **the use of ontogenetic development**, that is, studying functions not in phylogenesis but ontogenesis. Here there is less of the ready material, and we need to strain every nerve to reinforce this aspect of physiological studies because at present ontogenetic physiology is developed relatively poorly. We should try to embrace both the embryonic and postnatal stages of development and find out when the rudiments of different functions begin to show up, at what stage and how the function of those structures that pre-existed and developed independently changes depending on the formation of different morphological structures. This particularly concerns those tissues and organs that fall at a certain age during their development under the influence of the nervous system and endocrine factors. As a result of these effects, endocrine and neural, the developmental course itself undergoes substantial changes, and we should understand how the development would proceed without the interference of certain endocrine factors and the determinative influence of the nervous system on the development of endocrine organs as well as, vice versa, the effect of endocrine organs on the variability of the nervous system. With regard for all these features we should approach ontogenetic physiology to make it a tool for studying the evolution of functions.

Still, additional possibilities open up in front of us—**the use of clinical material**. I must say that it was not without reason that the issues of evolutionary physiology were raised by physicians, not biologists. This seems to be a paradox, but it is not true, and this is because clinical studies led to the idea that in certain cases some symptoms represent echoes of what had happened at earlier developmental stages, that in some cases of pathology, we deal with echoes of the evolutionary process, with reverting to those functional states that characterize the earlier developmental periods.

Of course, it is particularly easy to draw a comparison between clinical symptoms of different diseases and those phenomena that we observe during ontogenesis, but phylogenesis also reveals a lot in this respect.

Directly related to the use of the clinical material is **the use of special experimental methods**. These methods boil down to an artificial disconnection of individual organs and tissues from their controlling mechanisms, a disconnection inside the controlling mechanisms themselves, inside the nervous system, and a disconnection of some lower levels from the higher ones. Then follows an observation of changes that occur both in the lower parts of the nervous system depending on the loss of regulatory influences from the higher parts and, vice versa, in the higher parts as a result of the loss of those afferentations that arrive from the lower parts. In this respect, we are particularly lucky because we can afford carrying out any experimental transections in animals providing thereby new conditions for functioning of organs, tissues and parts of the central nervous system and endocrine apparatuses, which change their regulatory function under the influence of these disconnections. Hence, we can compare these data with the results of clinical pathology and those of comparative and ontogenetic physiology.

Based on these four methods of investigation, it is possible to get an idea of how the evolutionary process was proceeding and how functions were changing during their development. As a result, not only factual relationships (not merely descriptive though) would be established, but to a large extent it would be also possible to unravel the mechanisms of interaction and, hence, to a certain extent, resolve causal relationships as well. Still, that is not all. Experiments allow a special analysis of the effect of external factors. Now we have got an extremely abundant experience in the sense that modern science and technology have made it possible to generate such forms of energy that were unknown or unattainable previously. At present, they can be generated, detected, recorded and quantitatively evaluated. It is increasingly often that we are learning about the existence of those forms of energy that we have been unaware of before.

While throughout its evolutionary and historical development and until recently humankind was aware of only a limited number of energies that affected it naturally, over the last decades we have learnt that many types of energies represented in nature are much wider than we thought. Yet recently, ultrasounds seemed to be something artificially produced by people, but now it is turning out that during the evolutionary process not only were they generated by different animals but also served as means of signaling being perceived and evaluated in the same way as we perceive sound frequencies in the narrow range of audibility. Likewise, radio waves discovered by our compatriot Popov and being widely used in television and radio turn out to be generated by the Sun as well; the Sun's rays, that we assessed previously only in terms of their energy, contain also electromagnetic oscillations of those frequencies, wavelengths, that we use in radio engineering. So that over millions and billions of years during the evolutionary process, animal and human organisms were exposed to these electromagnetic waves, and it is only now, over just the last three decades, that we have come to grips with studying ultrahigh frequencies and applying them in ambulances and clinics for treatment, diagnostics, and so on. Nevertheless, they did exist in nature and affected all of us.

So far, we studied experimentally the effect of these ultrahigh frequencies for pure pragmatic purposes—their application in technology, medicine, and so on. Now they are becoming one of the possible factors of evolution, and we should set to studying ultrasound frequencies, electromagnetic waves and ultraviolet radiation not only in terms of their impact on individual functions and organisms but also in terms of their possible role in the evolutionary process. Thereby, we should look into the influence they may have on future generations.

If we investigate varied types of energies that we receive in their natural form and that we can now generate artificially, graduate, record and evaluate quantitatively, if we find out how they are reflected in the development of different functions, then we will obtain an enormous material for understanding not only the developmental course of functions and the history of the emergence of functional relationships but also their dependence on the environmental factors. Thus, we are arriving at a proposition that evolutionary physiology, as we understand it, should encompass a wide range of studies. It will certainly be interwoven with classical, applied and comparative physiology but anyway underlain with an endeavor to comprehend the causal dependence of the developmental course of functions on the external and internal factors, to understand those major lines, along which the evolution of functions is running and which jointly led the evolutionary process to proceed exactly in the way it did, and finally, to understand the diversity of pathways the evolutionary process may proceed along depending on those conditions that will be created on our planet. Here are the main tasks and main methods of evolutionary physiology.

However, the question arises whether the abovementioned exhausts the matter. Of course not. The basic principle, which underlies the evolutionary doctrine and studies carried out by the coryphaei of our physiological science, especially I.P. Pavlov's, is that organisms undergo the process of continuous adaptation to the environment. Therefore, it is not enough just to

http://dx.doi.org/10.5772/intechopen.73670

understand what specifically, and under what influence, is happening. We should clarify the adaptive role of the evolution of functions, understand how life was preserved and how it assumes various forms depending on the adaptation of a living being to new conditions under the influence of the external and internal environmental factors. We ought to understand which conditions are disastrous and which prove to be surmounted or secured by certain adaptations.

Elucidation of a number of adaptive mechanisms, adaptive changes in functions should again be one of the major tasks of evolutionary physiology. In this sense, evolutionary physiology will be not only a theoretical, but also strictly practical, science, since it will lead to results, which will enable us to influence the course of evolution in the future. It is quite important for medicine and zootechnics.

Especially hard questions arise when we approach a study of the human organism. We know well that the human organism at a certain stage of evolutionary development stopped to be only a biological being and became a social being. Interrelationships among people led to the establishment of certain social relations. The latter resulted from a kind of leap, or what may seem to be a leap, because some of our immediate predecessors perished and are now inaccessible for being studied. But in any case, the fact is that humankind rose above the rest of the animal world and in many respects surpassed it, creating new forms of activity as well as new forms of relationships with the environment. Man to a certain extent became the master of nature, at least of some of its aspects; he can consciously control it being expected to intensify his activity in this direction. Humans entered into certain relationships that are lacking in other animals. Man is not only a biological, but a social being, and this, on the one hand, relates to his nervous system development and transition to other forms of existence, but, on the other hand, these new forms of existence and interaction prove to be a powerful factor, which influences the course of changes in the structure and functions of an organism.

From this viewpoint, the historical period of the human existence certainly represents an extremely important stage in the evolutionary process, and evolutionary physiology should not digress from this point.

Recently, it seemed that the relationships among people, who created social living conditions and stay under the influence of these social factors, should serve a borderline, at which a physiologist must stop. However, proceeding from the I.P. Pavlov's doctrine and Marxist-Leninist theory, we can afford asserting that physiology must not stop here. The whole human organism with all its manifestations should become a subject of physiological investigation.

This does not mean that we should reject the existence of psychology and some other humanities. Quite the contrary, our task is to tie physiological studies to psychological as closely as possible in order to understand those physiological mechanisms that provided humans with the potential to turn from pure biological beings to simultaneously biological and social beings, to understand those physiological mechanisms that provided the possibility of interrelationships among people and thereby made this possibility a factor consciously directing the development of our progeny.

If we abandon this task, then both medicine in its major part, pedagogy and art will be absolutely cut off, ejected from natural sciences and left beyond the scope of natural-scientific investigation. This is not to be understood as an attempt to account for all social relations in terms of natural sciences. Of course, this would be incorrect, but providing a physiological basis for them should be our goal. No doubt, an exclusively important role in this respect was played by the I.P. Pavlov's doctrine about two signaling systems. If it were not for the two signaling systems, if the second signaling system were not superstructured over the first signaling system common to the animal kingdom, if those multiform influences that we widely use in fostering our children and interrelating with each other were not established, then we would have been ignorant of functional interconnections in the human organism and its relationship with the environment.

From this viewpoint, the aspiration for studying the human higher nervous activity during its formation and development is supposed to be the acme of evolutionary physiology. A comparison of the human and animal organisms shows a striking divergence in their developmental paths from the first hours of life.

When studying the postnatal development of animals and comparing it with that of human organisms, we see how considerably the range of possibilities for the nervous system development expands under the influence of the second signaling system, while the biological processes themselves become largely subjected to the influence of the second signaling system.

This stage of the evolutionary physiology formation should certainly interest us first of all, because it leads to the cognition of humankind while representing a junction between natural and social factors, which determine the development of human personality and its activity.

This is how I and a narrow circle of my fellow workers envision the tasks and methods of evolutionary physiology. I know that many of the abovementioned propositions are far from being new as they are recognized and being used for a long time. May be much of the said will prove to be worded unsatisfactorily. Anyway, underestimation of some of the points I have afforded to draw your attention to might have an unfavorable effect on the progress of evolutionary physiology.

Evolutionary physiology, as stated above, is supposed to be pivotal to independent research. It demands to enlist a number of experimental methods from physiologists, biochemists, morphologists and psychologists to create by joint efforts a science that would complement evolutionary morphology and, at the same time, provide a clear and comprehensive understanding of how the interrelationships among people and nature as well as among people themselves first formed and then altered during evolution under the effect of historic conditions. Simultaneously, such a high goal will be promoting the resolution of a whole number of practical tasks, because all our efforts to comprehend the causal dependence among functions of an organism as well as between the effect of external factors and activity of an organism will be a means of practical assistance to the population of our motherland and the whole world in defending against some adverse factors, which are present in nature, which we generate artificially and which expand their use. Without the knowledge of their role, we may prove to be helpless in combating pernicious influences. Such a science, essentially theoretical and being of profound practical importance, should become the goal of evolutionary physiologists.

http://dx.doi.org/10.5772/intechopen.73670

I reiterate that nowhere are the interests of theory and practice interwoven that closely as in evolutionary physiology. Not only will we be aware of how to struggle against the influence of some external factors on an organism but will obtain an enormous material to extend the range of our theoretical ideas from analyzing those conditions, under which the work is in progress at various scientific and manufacturing institutions. This unity of theory and practice should always underlie our approach to a problem and be a guiding star, allowing us not to shut ourselves off from the life practice and not to be afraid that one or another experimental path has a character of applied science. There is neither a theoretical nor practical science, there is a single science and there must be a single science. Practice must help theory and vice versa.